SpringerBriefs in Business

SpringerBriefs present concise summaries of cutting-edge research and practical applications across a wide spectrum of fields. Featuring compact volumes of 50 to 125 pages, the series covers a range of content from professional to academic. Typical topics might include:

- A timely report of state-of-the art analytical techniques
- A bridge between new research results, as published in journal articles, and a contextual literature review
- A snapshot of a hot or emerging topic
- An in-depth case study or clinical example
- A presentation of core concepts that students must understand in order to make independent contributions

SpringerBriefs in Business showcase emerging theory, empirical research, and practical application in management, finance, entrepreneurship, marketing, operations research, and related fields, from a global author community.

Briefs are characterized by fast, global electronic dissemination, standard publishing contracts, standardized manuscript preparation and formatting guidelines, and expedited production schedules.

Carol Aebi

Unifying Sustainability Information for Societal Automation

Leveraging Collective Intelligence

 Springer

Carol Aebi
Business Science Institute
Wiltz, Luxembourg

ISSN 2191-5482   ISSN 2191-5490   (electronic)
SpringerBriefs in Business
ISBN 978-3-031-83119-5   ISBN 978-3-031-83120-1   (eBook)
https://doi.org/10.1007/978-3-031-83120-1

This work was supported by Carol Aebi.

This Springer imprint is published by the registered company Springer Nature Switzerland AG
The registered company address is: Gewerbestrasse 11, 6330 Cham, Switzerland

If disposing of this product, please recycle the paper.

I release these words to the world, may the ideas contribute to something useful.

Preface

We find ourselves at a juncture; the fusion of business, technology, and society is propelling us at an unprecedented speed into the future. This convergence presents us with unparalleled opportunities but raises concerns about its potential unforeseen impacts on society. As machine learning and artificial intelligence advance, developing solutions prioritizing societal sustainability becomes increasingly urgent, specifically integrating individuals' well-being. Widespread concern about the impact of technology has sparked various views and global conversations, many echoing dystopian visions like Orwell's. Yet, as the pace of change accelerates, many businesses, governments, and institutions worldwide have little tools other than attempting to balance economic success with social responsibility through regulation. However, there is another perspective—synergistic alignment embedded within the technological algorithms. I am inspired by technological advances and see them as an extraordinary opportunity to act as a catalyst for positive change, fostering engagement and contributing to realizing sustainable and harmonious outcomes.

I believe that stakeholder collaboration, data integration, and automation are essential to social sustainability. The key lies in handling vast data sets across scales, now achievable through technology. This belief inspired my doctoral research on building a stigmergic information systems network for social sustainability, a concept designed to unify the vast streams of data and information collected through sustainability assessments to support intelligent, decentralized decision-making for societal good. Stigmergy, a term borrowed from biology, refers to the indirect coordination of actions through environmental signals. In the context of sustainability, this concept provides a framework for automating the capture, measurement, and connection of data related to social value indicators, ultimately supporting informed decision-making at multiple levels of society. The idea is that we can choose to drop relevant pieces of information and data as we go about our daily "technology-driven" activities and train technology to pick up this relevant data and sort and apply it to prevent or solve problems for the betterment and synergistic alignment of society.

The book "Unifying Sustainability Information for Societal Automation: Leveraging Collective Intelligence for a Sustainable Future" builds upon my doctoral research and translates the insights from that research into an accessible yet comprehensive exploration of how a stigmergic information system network can serve as a model for social sustainability. Whether examining the micro level of individual behaviors, the meso level of organizational actions, or the macro level of societal policies, this model provides a first step in a unifying pathway for integrating diverse forms of data in ways that align with evolving social needs.

I am a business executive and researcher with over two decades in technology, consulting, and governance, focusing on business strategy, digital transformation, business analysis, and corporate governance. My experience advising corporate, SME, and start-up businesses as a leader, board member, and consultant has fueled my passion for aligning business needs with technological solutions that benefit the private sector and society.

I hold an Executive Doctorate in Business Administration, and this book is based on my 2022 doctoral thesis, *A Conceptual Model of a Stigmergic Information System Network for Social Sustainability: An Inductive Top-down Theorizing Approach*. This research resulted from an international collaboration between the Business Science Institute in Luxembourg, SKEMA Business School in France, Université Jean Moulin Lyon 3 in France, and the University of Technology Sydney in Australia, under the thesis supervision of Prof. Lapo Mola of the Skema Business School, Digitalization Academy Knowledge, Technology and Organization Research Center in Sophia Antipolis. This doctoral research was rigorously examined by an international panel by Prof. Paul Beaulieu, professor at the Environmental, Social, and Governance (ESG) of the University of Quebec in Montreal, Canada, and Prof. Elaine Ferneley, Director of Keele Business School at Keele University, ultimately approved by a jury presided over by Prof. Christophe Elie-Dit-Cosaque, Professor of Management Sciences and Information Systems at Université des Antilles.

This book provides a fundamental and unifying perspective across disciplines, offering a model for social value that adapts to technological and societal changes. The model presented here is designed to be a foundation where additional algorithms and decision branches are expected to be developed, evolve, and adapt as society changes, offering a pragmatic solution to the dynamic challenges we face today. I also view it as a starting point—an open invitation for others to contribute to this evolving field, refine further, and build upon the ideas presented. It is both a theoretical concept and a practical tool for navigating the often-fragmented world of social sustainability metrics and strategies.

As a researcher and business leader, I have always believed that the most effective solutions arise from interdisciplinary collaboration and the synthesis of diverse perspectives. I hope this book serves as both a guide and an inspiration for those seeking to navigate the complexities of sustainability in a data-driven, digitally enhanced world. By connecting technology with societal needs and data with human values, it is my dream that we can unlock new opportunities for sustainable systems and positive impact. With this belief, I present this book and invite you to join me in exploring its possibilities.

Wiltz, Luxembourg   Carol Aebi
September 2024

Acknowledgments

This work is the product of several minds, hands, and hearts, and I am deeply grateful to everyone who has contributed to its completion.

First and foremost, I would like to express my profound thanks to the Springer Publishing Team for the meticulous care you put into making this book the best it could be.

To my supervising Professor Lapo Mola of SKEMA Business School, I owe a debt of gratitude for your intellectual rigor, your constructive challenges, and your constant insistence on clarity and precision. Your guidance helped me navigate the complexities of this subject in my attempts to distil it into a coherent and compelling narrative.

A special thank you goes to my partner, Johannes, whose support has been invaluable in bringing this book to publication.

A heartfelt thank you also goes to my "F"riend Bob Boylan, who took the time to read the original thesis draft in its messiest form and provided valuable editing comments. I deeply appreciate your thoroughness and generosity of time and effort.

I would like to express my gratitude to the extended academic team and doctoral students across the partnering institutions that helped shape and support this research: Business Science Institute, SKEMA Business School, University of Technology Sydney, and Université Jean Moulin Lyon 3. Your comments sharpened my arguments and pushed me to think more deeply and clearly. Your collaboration, intellectual input, and support have been indispensable in making this work possible.

Finally, I extend my gratitude and thanks to Muhammad Ali Quresh. His wizardry in formatting, attention to detail, and remarkable expertise with LaTeX, and table formatting made a world of difference. And Bree Byle of Bree Rose Creative LLC for her beautiful graphic illustrations of the modeling, which brought a fresh visual dimension to this work.

To all of you—thank you for being part of this journey.

Contents

Acronyms

AI	Artificial intelligence
ACO	Ant Colony Optimization
ARIES	Artificial Intelligence for Environment & Sustainability
BER	Basic Efficiency Resource
BoP IAF	Base of the Pyramid Impact Assessment Framework
CA	Capabilities Approach
CATWOE	Customers, Actors, Transformation Process, Worldview, Owners, Environmental Constraints
CSR	Corporate Social Responsibility
EIA	Environmental Impact Assessment
ESG	Environment Social Governance
FIFA	Federation International de Football Association
GRI	Global reporting initiative
IAIA	International Association for Impact Assessment
IAP	The InterAcademy Partnership
IFC	International Finance Corporation
IFRS	International Financial Reporting Standards
IGO	Intergovernmental Organization
ISSB	International Sustainability Standards Board
ITDTA	Inductive Top-Down Theorizing Approach
KPI	Key Performance Indicator
MIF	Measuring Impact Framework
MIT	Massachusetts Institute of Technology
NGO	Non-governmental Organization
NPO	Non-profit Organization
OECD	Organization for Economic Co-operation and Development
OGP	Open Government Partnership
PIA	Participatory Impact Assessment
SDG	Sustainable Development Goals
SES	Social-Ecological System Framework
SI	Social Impact

SIA	Social Impact Assessment
SIS	Stigmergic information systems
SME	Small to Medium-Sized Enterprise
SROI	Social Return on Investment
SSM	Soft Systems Methodology
STI	Science, Technology and Innovation
SV	Social Value
SWB	Subjective Wellbeing
TFM	Technology Facilitation Mechanism
UML	Unified Modeling Language
UN	United Nations
UN-CEBD	UN Committee of Experts on Big Data and Data Science for Official Statistics
UN DESA	UN Department of Economic and Social Affairs
UNEP	United Nations Environment Program

List of Figures

List of Tables

Chapter 1
Introduction to Unifying Sustainability Information

1.1 Introduction

The advent of societal automation has emerged and is catalyzing pivotal discussions. Societal automation wields the potential to influence numerous aspects of daily life, leading to potentially controversial outcomes. Societal automation is a comprehensive subject encompassing communication system networks, computing paradigms, algorithms, data analytics, network sensors, architectural frameworks, systems and applications, smart cities, cyber-security, large-scale ultra-complex engineering, human behavior modeling, as well as monitoring, control, and assessment within adaptive, interactive societal systems. Careful consideration of societal automation technology can leverage collective intelligence to support sustainability decision-making.

This book focuses on an initial step in the endeavor to unify existing social sustainability assessment information considering the perspectives of governments, organizations, groups, and individual stakeholders, and in formulating a first-tier-level information system network model. This model serves as a guide for theorists, policymakers, technologists, activists, managers, and investors, facilitating the development of technologies that will interact, operate, and reason in accordance with societal expectations regarding social value. This foundational model may be further elaborated beyond data and information dissemination and eventually applied as a cornerstone for societal automation. This model harnesses the new phenomenon of scale-free distributed technology for social sustainability. It is structured to offer predictive and suggestive analysis to minimize negative social impact. Stigmergic self-organization processes capture data for information aggregation to form a collective intelligence to guide decision-makers toward activities that solve social problems. Embedding sustainability information within societal automation may add a wrapper of trust that balances the potential for heightened efficiency and productivity with the imperative of society to mitigate adverse social impacts, a crucial undertaking for policymakers, technologists, and society as a whole.

© The Author(s) 2025
C. Aebi, *Unifying Sustainability Information for Societal Automation*,
SpringerBriefs in Business, https://doi.org/10.1007/978-3-031-83120-1_1

In contemporary discussions, the role of business in society is experiencing considerable examination and transformation. The harmonization of business endeavors with social advantages, often encompassed under the concept of "shared value," has emerged as a focal point for various stakeholders. We are already deep within the era where the alignment of profit with social good is paramount, and understanding and measuring the societal impact of business activities has become a pressing concern for businesses, governments, and investors. The challenge lies not in the lack of metrics and assessment tools but in their overwhelming abundance, leading to fragmented and often incomparable data.

There is a growing interest in measuring and comparing social value and social impact results of businesses from various stakeholders. Enterprises increasingly acknowledge the need for strategic alignment with their communities [8]. Simultaneously, social organizations must exhibit their effects to secure funding and support from partners and beneficiaries [7]. Governments and regulatory bodies endorse integrated approaches to collect comparable data from businesses [2]. Funding bodies are also keen to invest in the most efficient and socially responsible projects [7]. Policymakers require comprehensive data to educate their constituents and substantiate spending decisions [7]. This widespread demand for comparable and integrated data necessitates an innovative social value and impact assessment approach.

This growing interest has led to a cluttered landscape of competing assessments, metrics, and tools [7]. Each method is tailored to the specific needs of individual organizations, governmental programs, or local communities, making comparative analysis challenging [8]. Standardized, comparable data is crucial for effective decision-making and policy formulation. However, achieving this standardization in a dynamic societal context is a formidable challenge.

The rise of ubiquitous IT and digitalization presents a promising solution to this issue and social automation in general. Technological progress can enhance the efficiency of data collection and assessment on a large scale. Insights from social learning and activity theory indicate that learning processes, often considered uniquely human and culturally influenced, can be computationally modeled [3, 4]. This shift towards "cognition in practice" in computational models creates new opportunities for data-informed decision-making. However, this technological potential brings its own set of challenges. Society is a continually evolving dynamic system, while computational systems, algorithms, and machines demand consistency and well-defined rules to operate effectively. The implications of embedding such rules and decision structures in computational models are extensive, potentially impacting policy and business decisions. Initially, these systems may function as mere data collection and data-sharing mechanisms, but they might also offer decision-making recommendations for social automation as they progress. Therefore, involving stakeholders in responsible programming ensures that these systems align with societal values and expectations.

What is urgently needed is an initial guide or model that allows stakeholders to evaluate the alignment of technological systems with societal values and requirements. Such a model should facilitate discussions among developers, computer scientists, social sustainability practitioners, business managers, and policymakers. It should provide a foundation for advancing both theoretical and practical aspects of social value assessment and integration into technological systems.

"Unifying Sustainability Information for Societal Automation: Leveraging Collective Intelligence" addresses this challenge by presenting a scientifically structured approach to harmonizing social sustainability data across various scales of participation. The book delves into model building, which integrates sustainability data, systems thinking, systems dynamics, and stigmergic systems. This book synthesizes existing literature through the Inductive Top-Down Theory Building Approach to construct a scale-able, automated framework that enhances decision-making and supports policy design. This model offers a unifying perspective on social value and impact assessment data contextualized within scale-free distributed information systems, which are information systems that are not dependent on a specific scale or size and are designed to operate across varying scales of societal complexity. The resulting model facilitates the systematic collection, analysis, and communication of data among stakeholders, ensuring that the societal impact of business activities is both measurable and manageable.

Four theoretical pillars underpin this research: sustainability, systems thinking, systems dynamics, and strategic systems. This book applies the definition of sustainability as "the capacity to endure," encompassing environmental and social dimensions [5]. Social sustainability, specifically, refers to the capacity of human societies to maintain stability and endure over time. This concept considers frameworks such as the United Nations Sustainable Development Goals (SDGs) and Raworth's social and planetary boundaries. Systems Thinking involves analytical skills to understand and model social systems, predict behaviors, and devise solutions for complex social challenges [9]. It is essential for building models that can automate workflows and influence desired outcomes. Systems Dynamics provides tools for analyzing complex societal activities, including stocks, flows, internal feedback loops, and time delays [1]. This approach introduces predictive capabilities crucial for policy design and continuous adaptation. Stigmergic Systems refer to mechanisms that facilitate self-organization through indirect coordination [6]. In the context of information systems, stigmergic systems offer a framework for organizing software services based on activity fields, enabling scale-able and flexible data management.

The research adopts the Inductive Top-Down Theory Building Approach (ITDTA) by [10]. This methodology is particularly suited for uncovering new phenomena within a complex body of existing knowledge. It supports discovery and theory development by combining literature review, pattern recognition, and contextualization. This research stems from the September 2022 Doctorate in Business Administration thesis by Carol Aebi within a multi-university program delivered by the Business Science Institute, SKEMA Business School, and IAE Lyon School of Management. Supervised by Prof. Lapo Mola of SKEMA Business

School and reviewed by distinguished professors from ESG UQAM and Keele Business School, this research presents both an academic as well as managerial contribution to social sustainability and information systems.

The book chapters are organized in a format that reflects the conceptual theory-building process, which applies building blocks. Chapter 2: titled The Who, What, When, Where, and Why of Social Impact Data is the first building block which takes a structured view of the existing literature on social impact and sustainability assessment literature uncovering the relationships between the different stakeholders that are assessing social impact today and sorting the differences that can be applied as requirements, a critical starting point for the model's construction. Chapter 3: titled Metaphor-Driven Contextualization: Reordering Sustainability Information for Collective Intelligence is the second building block. It involves thought trials through the use of metaphor, specifically leveraging the well-known metaphor of Flux and Transformation by Gareth Morgan to guide the exploration of the four logics of change: Relationships between Systems and their Environments, Chaos and Complexity Theory, Cybernetics; and Change as the Tension between Opposites. This process contextualizes the new phenomena of scale-free distributed technologies within an environment of dynamic societal complexity to offer novel insights for unifying the various perspectives. Chapter 4: titled From Concepts to Coherence: Modeling Social Impact Data for Systemic Alignment is the final building block introducing the conceptual model that aligns the requirements identified in the first building block with the conditions uncovered in the second building block, integrating pragmatic technical components for systemic alignment and the emerging results. The fourth chapter concludes with research implications and recommends further studies. A list of abbreviations and glossary references are available in the last sections.

The emerging result is a Conceptual Model for a Stigmergic Information System Network for Social Sustainability, which offers a groundbreaking framework for unifying social value and impact assessment data. It provides a new contextualization for applying these assessments within distributed information systems, capable of operating across various societal scales including the individual. It further incorporates predictive, corrective, and retrospective analytics which function as a systemic thermostat for social stability, reducing perception bi-as, and monitoring exploitative effects.

This research lays the foundation for further research and elaboration, ultimately contributing to a more sustainable future with societal automation that supports decision-making more closely with societal value. Embark on a journey with this groundbreaking exploration of technology's emerging role in societal impact and value alignment. Discover how this unified model bridges gaps across disciplines, providing a stable foundation amidst ever-changing societal landscapes and paving the way for a sustainable future.

References

1. Arnold RD, Wade JP (2015) A definition of systems thinking: a systems approach. Proc Comput Sci 44:669–678
2. Business Call to Action (2016) Measuring impact: how business accelerates the sustainable development goals. United Nations Development Programme and GRI
3. Engeström Y (2015) Learning by expanding. Cambridge University Press, Cambridge
4. Lave J (1988) Cognition in practice: mind, mathematics and culture in everyday life. Cambridge University Press, Cambridge
5. Liu S (2020) Bioprocess engineering: kinetics, sustainability, and reactor design. Elsevier, Amsterdam
6. Marsh L, Onof C (2008) Stigmergic epistemology, stigmergic cognition. Cognit Syst Res 9(1–2):136–149
7. Mulgan G (2010) Measuring social value. Stanford Soc Innov Rev 8(3):38–43
8. Porter ME, Hills G, Pfitzer M, Patscheke S, Hawkins E (2012) Measuring shared value: how to unlock value by linking business and social results. Foundation Strategy Group (FSG), (46910)
9. Richmond B (1994) System dynamics/systems thinking: let's just get on with it. Syst Dyn Rev 10(2–3):135–157
10. Shepherd DA, Sutcliffe KM (2011) Inductive top-down theorizing: a source of new theories of organization. Acad Manage Rev 36(2):361–380

Chapter 2
The Who, What, When, Where and Why of Social Impact Data

2.1 The Interconnection of Social Impact and Sustainability

Social impact and sustainability are interwoven constructs that collectively describe the trajectory of our society, economy, and environment. Fundamentally, both seek to conceptually express and, therefore, cultivate goals that foster a world where the present needs are met without compromising the ability of future generations to meet their own needs. While sustainability primarily focuses on the long-term viability of environmental, social, and economic systems, social impact concerns the effects of actions, policies, and practices on individuals and communities. Social impact and social value are terms frequently employed interchangeably to denote the beneficial or detrimental effects of activities.

Sustainability is often viewed through the prism of environmental stewardship, but it also encompasses social and economic dimensions. In this research, sustainability is defined broadly as the capacity to endure [27]. Social sustainability is defined as the capacity of human society to survive. Social sustainability refers to the positive and negative impacts of formal and informal processes, systems, and relationships for stable communities and future generations. For a society to endure is to navigate a continuous journey of existence, preserving through the ages and ensuring that it endures beyond the bounds of time.

Until recently, social sustainability was neglected as a dimension related specifically to sustainable development at a community, state, or country level [26, 29, 30]. Sustainable development considers inter-generational equity, focusing on meeting the needs of present generations without compromising the ability of future generations to meet their own needs [4].

To explain why this research applies a broad term for sustainability, please consider that sustainability alone is more general than sustainable development at a regional, global or governmental level. Sustainability refers to processes, objects, or matters where competing forces exist, such as within a society, organization, or family. Resourse usage in one area may lead to exhaustion or depletion of the growth

© The Author(s) 2025
C. Aebi, *Unifying Sustainability Information for Societal Automation*,
SpringerBriefs in Business, https://doi.org/10.1007/978-3-031-83120-1_2

source. The term sustainability is compatible with an overall state of stability, the stability of an individual, the stable state of an organization, or the stability of a society [27].

This research applies the principles of sustainability and sustainable development to society (macro level) while also considering sustainability at a group or organization (meso level) and individuals (micro level) due to the vastly different perspectives that require different data classifications and processing by information systems. Therefore, the relationship between social impact and sustainability is symbiotic. Effective sustainability practices inherently produce positive social impacts, while meaningful social impact initiatives support broader sustainability goals.

2.2   The Promise of Enhancing Decision-Making Through Assessment, Data, and Information Sharing

In computing, data is integral to feeding processes and operations. It serves as the fundamental input, delineating the foundation upon which all subsequent processes are selected. Information can be an input or an output. As input, we provide data, instructions, and commands to guide the processing of other data. As output, it represents the culmination of processing and interpreting data. It encompasses the results of our queries and commands, presented in a form we can comprehend and act upon. In this context, information refers to what computing systems share with other systems or provide back to us after processing the data.

When we think about sustainability, social impacts, and automation, we consider what data is being collected, categorized, interpreted, and how that information will be used for different purposes by different people or organizations. Data quality and standardization is essential for machine supported decision-making in societal automation. Unification and synchronization of data can enable information dissemination for positive societal and environmental change while safeguarding against potential negative consequences.

Currently, organizations have a myriad of disparate and heterogeneous measurement toolboxes for assessing and evaluating social sustainability, making it challenging to standardize and harmonize measurement practices across the board. There is a lack of consensus on the definition of social impact, and congruent measurement hampers the applicability, comparability, and academic debate on social impact and the usage of consistent methods [6, 28–30, 34, 35, 42, 45]. Moreover, concepts, hierarchies, and categorizations are also currently inconsistent [28].

Social sustainability is often evaluated through a Social Impact Assessment (SIA), which examines the effects of projects, policies, and initiatives on social well-being and equity. SIA involves assessing, analyzing, monitoring, and managing the intended and unintended social consequences, both positive and negative, of planned interventions such as policies, programs and plans [22].

The Sustainable Development Goals (SDGs) established by the United Nations offer a comprehensive action plan for a global sustainable development partnership to enhance lives and safeguard the environment [43]. The SDGs are centered on sustainable development, as articulated in the Brundtland report Our Common Future by the World Commission on Environment and Development. According to [4], sustainable development is defined as "development that meets the needs of the present without compromising the ability of future generations to meet their own needs" (p. 6). This emphasis on present and future well-being urges policymakers to adopt a risk perspective that prioritizes stabilizing factors while avoiding destabilizing ones. It is important to note that social sustainability is not the sole element of stability, as sustainability is also interconnected with environmental resources and planetary well-being.

Society is a system of interconnected parts that cooperate [15]. Social sustainability considers the viability of the endurance of society. When everything and everyone is reliant upon something else, or someone else, for their existence, this gets complicated. As a basic example, people need oxygen, water, nutrition, and rest to sustain their bodies, but they also want, some say need, love, companionship, purpose, education, healthcare, and meaning. Organizations have needs, such as funding, workers, customers, products, and resources. Communities require infrastructure, governance, and order, etc.

Not so long ago, there was a belief that measuring and monitoring environmental sustainability was too complex to tackle. Fast forward to today, and we have made significant progress. Assessment for ecological sustainability has advanced to an ever-improving common language and concepts. The way we measure and monitor environmental sustainability has come a long way. In 2021, the UN Department of Economic and Social Affairs (UN DESA) and the UN Environment Program (UNEP) launched the Artificial Intelligence for Environment & Sustainability (ARIES) for the System of Environmental-Economic Accounting (SEEA). This digital tool, available on the UN Global Platform, allows businesses to simultaneously track their environmental impact and economic performance. ARIES represents a significant step in using AI to advance environmentally Sustainable Development Goals (SDGs). This progress in tracking environmental sustainability metrics gives hope that similar complexities in social sustainability can also be addressed with new tools and approaches.

One researcher who is addressing the complexity of social sustainability information in a societal sustainability context is Dr. Kate Raworth. Her research on human and societal development has introduced the concept of the "Doughnut," which represents the social and planetary boundaries within which humanity can thrive without causing environmental degradation. This concept combines the nine planetary environmental boundaries defined by Rockstrom et al. with twelve dimensions derived from the Sustainable Development Goals (SDGs) of 2015. The planetary ecological boundaries set out by Rockstrom et al. establish the limits for acceptable environmental degradation of Earth's ecosystems [37]. Raworth argues that operating within these boundaries will create a safe and just space for humanity to flourish. Furthermore, these boundary conditions provide a basis for algorithmic computation at a macro level, offering a data-focused framework for analysis and decision-making [36]. Projects are being launched to begin collecting national datasets that align with the Raworth Doughnut. A future where these national boundary conditions are applied within models for societal automation is increasingly possible.

Businesses and organizations are obligated to shareholders to focus on profits that ensure their sustainability. This focused perspective can lead to repercussions on society in the form of products, services, jobs, working conditions, human rights, health, innovation, environment, and education [17]. Many businesses are investing in understanding their social impacts to harmonize their efforts with the goals of the societies they operate within. However, the complex and costly nature of assessing social sustainability presents considerable business challenges. Moreover, the lack of comparability of data resulting from divergent perspectives and methodological approaches hinders informed business decision-making. These inconsistencies give rise to various risks, such as the potential for false precision, reliance on a singular numeric metric to gauge investment success, and the risk of cherry-picking. These risks may result in overlooking the negative impacts on the most vulnerable populations [42].

As the urgency of addressing sustainability and social impact grows, technology and automation have emerged as powerful tools to advance these goals. Innovations in data analytics, artificial intelligence, and digital platforms are transforming how we assess, manage, and enhance social and environmental outcomes. Automation, for example, facilitates real-time monitoring of resource usage, emissions, and social metrics, enabling more precise and timely interventions. Furthermore, advanced data systems can consolidate extensive social impact data, unveiling previously inaccessible patterns and insights. These technological advancements facilitate the implementation of more efficient, scale-able solutions that align with sustainability goals while maximizing positive social outcomes.

The integration of automation technology is increasingly important in unifying fragmented sustainability data. Automation significantly stimulates this unification of definitions, categorizations, classifications, and use cases for information sharing by demanding clarity and consistency in content during the programming and modeling processes. Automating complex tasks such as data collection, analysis, and reporting offers cost-effective options to ensure that sustainability factors are consistently tracked. This, in turn, facilitates comparison and transparency. Furthermore, it offers the potential for enhancing accountability and enabling more informed decision-making by including more comprehensive data.

According to Deming, a system must be managed. It will not manage itself. Left to themselves, system components become selfish, competitive, independent profit centers, thus destroying the system. The secret is cooperation between components toward the aim of the organization. We cannot afford the destructive effect of competition [10]. Societal automation offers an optimistic view of the future where information systems share and transforms broad assessment data into usable information to direct competitive forces of business toward rapid solutions to social problems within sustainable boundary conditions. Integrating technology and automation into the realm of sustainability and social impact is not just an enhancement but potentially a necessary evolution to embed the probability of trust within the system.

What gets measured gets managed [14]. So, what gets measured for social sustainability? Social value and impact are two overarching terms that measure individual social sustainability variables. Many social value theories and approaches exist across cultures, industries, and markets. This presence implies both a demand as well as complexity.

Social impact assessment (SIA) arose alongside the environmental impact assessment (EIA) as a sustainability metric by the U.S. National Environmental Policy Act (NEPA) in 1970 [45]. Relevant literature on organizational effectiveness peaked between 1975 and 1985, with a potentially infinite number of models [26]. By the mid-2000s, the United Nations (UN) initiated international reporting approaches. Since then, business academics and proactive companies have called for collaboration to build consensus around the question of how to assess the positive and negative social impacts created across organizations, businesses, and projects [19, 28, 34, 39].

2.3 An Investigation into Social Value and Social Impact Assessment

Building Block I takes a structured exploratory dive into social value / social impact literature. Theories, metrics, assessments, and tools that specifically document measuring social value for business and or societal institutions were analyzed. Studies were considered if the business measurements were general and not industry-specific. Private and public sector businesses were categorized as "business." National governments, as well as organizations measuring social value within non-governmental organizations (NGO) or non-profit organizations (NPO) and intergovernmental organizations (IGO), were included within the categorization of "institution." Only research that took a social value view from a general national, country, or global societal perspective were included, e.g., the United Nations (UN), Organization for Economic Co-operation and Development (OECD). Single community-specific, culture-focused, or socio-economic-level bound studies were excluded. To refine and focus the topic, similar research content was also excluded: Psychology of social value, philosophy of social value, ethics of social value, social capital, and exclusively investment-specific measurement. Empirical as well as conceptual studies were included. No restrictions were applied on conference articles, journal rankings, research design, source of funding, or authors.

Search terms included: "social value" and "measurement," "shared value" and "measurement," "social impact" and "measurement," "social value" and "assessment," "shared value" and "assessment," "social impact" and "assessment." The search was conducted in the Title, abstract, and keywords. Reference lists were examined.

The following databases were searched: EBSCO database, Shared Value Initiative resource portal, Social Value Portal, World Bank eLibrary, United Nations resources, Open Government Partnership resources, Google Scholar web search engine, OECD digital library, ProQuest database, JSTOR database, Science Direct database, and Emerald Insights database.

Seventy-three documents were identified for screening. Fifteen studies met the criteria for inclusion in the review.

Data extraction followed guidelines by Kitchenham and Charters [25]. Data extraction from each study included five sections for each selected research. The general section documented the objective of each study and research question, history, future work suggestions, and the challenges for measuring social value. The Social value terminology section documented terms used, definitions, synonyms applied and references. The Research design section documented research epistemology and research methods. The Research theory section documented why the research was measuring social "x," the methods for measuring social "x," the view of business' role, and the conclusion of the research. The Boundary conditions section documented if the research was empirical or conceptual, the subjects: sample size and demographics if applicable, the Stakeholders: business, institutional, or both, the sustainability association, the level of research, e.g., national or international society, as well as the date of each study.

The social value/social impact literature review applied an exploratory synthesis approach. This identified stakeholder patterns such as relationships, purpose, and challenges for measuring social value/social impact by business and institutional stakeholders. Stakeholders, assessment methods, and purpose were also mapped as a visual tool for understanding the interrelationships and how stakeholders are connected within the literature. Predominant assessment methods were mapped across the literature. The purpose of assessments was to explore the unique perspectives of different stakeholders. Similarities and variations of purpose, definitions, and reasons for measuring social value documented in the existing literature were analyzed. The exploratory review gained an understanding of the challenges, the purpose of assessment, terminology, and definitions from different stakeholder perspectives. Output from Building Block I included relationship mapping and requirements gathering, which contributed to conceptual modeling in Building Block II and Building Block III.

The protocol followed the systematic review guidelines by Kitchenham and Charters [25] to enhance the quality and rigor of the literature review. The protocol includes training, practical screening, search, data extraction, and synthesis.

2.3.1 Distribution of Stakeholders

The literature review study considered two general classifications of stakeholders: businesses and institutions. Literature that considers social value views from a generalizable national, country, or global societal perspective is included, e.g., the United Nations (UN), and the Organization for Economic Co-operation and Development (OECD).

The business social value assessment literature included private and public sector approaches that are generalizable and not industry-specific.

The institutions' category included assessments for national governments as well as organizations measuring social value within non-governmental organizations (NGO) or non-profit organizations (NPO) and intergovernmental organizations (IGO). Multilateral initiatives such as the Open Government Partnership (OGP) and interdisciplinary academic research institutions are also included in the category of "institution." It is understood that these classifications are general and may be segmented or reclassified as the theory emerges.

13% of the studies focused on the perspective of the social value/social impact assessment from the perspective of Institutions. 40% of the studies focused on the perspective of the social value/social impact assessment from the primary perspective of Business. Almost half of the studies selected, 47%, consider both a business and institutional perspective for social value assessment.

This strong representation of the combination of business and institutional perspectives supports the hypothesis that relationships and inter-dependencies exist between these primary stakeholder groups when assessing social value/social impact.

2.3.2 Value or Impact: The Terminological Ambiguity that Technology Cannot Handle

The terminology, concepts, and expected results for social value in the social value and social impact assessment literature are inconsistent. This inconsistency poses a dilemma in assessing the dynamic and complex variables of social sustainability.

There is broad agreement across the studies that the lack of agreement on the definition of social impact and congruent measurement hampers the applicability, comparability, and academic debate on social impact, as well as the usage of consistent methods [6, 28–30, 34, 35, 42, 45].

Social value and impact assessments are currently performed for specific projects, companies, or governments. The author has the flexibility to present conclusions of social value or social impact in a unique manner.

Table 2.1 list and sort the primary term of what is assessed (social value or social impact), its stated definition, source of the definition, and synonyms in the literature sample for this research. All studies agree that Social Impact concerns the resulting

Table 2.1 Social value and social impact literature study definition review (1 of 2)

Citation	Stake-holder view		Terminology			
	Business	Institutions	Term used	Definition	Synonyms	Definition reference
Business call to action [6]	X	X	Social impact	Sustainability reporting and impact measurement are two practices used by companies to improve performance, account for impact, and publicly communicate sustainability data. Companies featured in this report use sustainability reporting and/o impact measurement to define, measure and monitor their social, economic, environmental and governance performance (called 'impact' in this report).	Sustainability reporting	N/A
Colantonio [8]		X	Social sustainability	There is general agreement that the different dimensions of sustainable development (e.g., social, economic, environmental, and institutional) have not been equally prioritized by policymakers within the sustainability discourse.	N/A	[2, 13, 20]
Domínguez et al. [12]	X	X	Social factors	N/A	Social impact, social aspects, the need to take into account ethical and political diversity in the socio-environmental	SIA / EIA
Goedkoop et al. [19]	X		Social impact	A broad range of social topics (24, see page 25) including business dependencies and social impacts	N/A	N/A
Kato et al. [24]	X	X	Social value	There is a lack of consensus on the definition of social value.	Social Impact	[28]
Kroeger et al. [26]	X		Social value	Not defined	Social value	N/A

Source			Term	Definition	Note	References
Maas et al. [28]	X	X	Social impact	By impact, we mean the portion of the total outcome that happened as a result of the activity of an organization, above and beyond what would have happened anyway.	The term social impact is often replaced by terms such as 'social value creation' and 'social return'	[7, 16]
McGinnis et al. [29]	X	X	Social	Overarching term that can be applied to Social theories, social choice theory, social dilemmas, social constructivism, social-political-economic settings, social variables, social collaborations, social-technical systems, social-ecological systems, social performance measures	See definition	[3, 5, 21, 31, 32]
Mulgan [30]	X	X	Social value	Unfortunately, there is no single authoritative definition of "social value," but we can say that it refers to wider non-financial impacts of programs, organizations, and interventions, including the wellbeing of individuals and communities, social capital and the environment.	Social Impact	[30]
Polonsky et al. [33]	X		Social value	The non-monetary benefits to wellbeing across stakeholders such as beneficiaries, society, donors, etc.	We use the terms social impact and social value interchangeably, where social impact is defined as the total impact on all its stakeholders	[9, 11, 33]
Porter et al. [34]	X		Shared value	Shared value separates social value and social impact. A way forward for integrated reporting to recognize the difference between what is measured to demonstrate impact and what is measured to capture value creation	Social value	N/A
Potma [35]	X		Social impact	Social impact is the share of the total outcome that occurred as a consequence of the activity of a company, above and beyond what would have happened anyway.	Social value created (SVC) "is the positive change in the social well-being (SWB) of disadvantaged individuals, caused by a social intervention"	[7, 26]

(continued)

Table 2.1 (continued)

Citation	Stake-holder view		Terminology		Synonyms	Definition reference
	Business	Institutions	Term used	Definition		
Tuan [42]	X	X	Social value	We will use the term —social value creation ‖ or —social value ‖ throughout the paper to refer to the general concept and practice of measuring social impacts, outcomes and outputs through the lens of cost.	Social impact: note a clear differentiation with Social Impact from the definition of "Impacts"	N/A
Vanclay et al. [45]	X	X	Social impact	Something that is experienced or felt, in a perceptual or corporeal sense at the level of an individual, social unit (family/household/collectivity) or community/society. (also see Social change process)	Shared value focuses on addressing societal needs alongside economic goals, redefining a company's purpose to benefit society and the business. It highlights that social issues can create costs or risks for firms, which must be managed for long-term success.	[44]
World business council for sustainable development [46]	X		Social impact	Impacts on society	N/A	N/A

consequences to society associated with a change in a population or group over time. When assessing social value, the term "social impact" is most frequently used as a primary term. However, social impact is also documented as a synonym of social value in 13 of the 15 studies. In 6 of the 15 studies, social value and social impact are used interchangeably throughout the text, often with only a subtle difference.

Research studies and business reports may be able to handle this level of terminological ambiguity. However, for computers to sort data and apply the information within algorithms, clarity of terminology is required to ensure data quality and comparability.

A Building Block I Requirement is to clarify the primary terms for quality cross-disciplinary data integration.

2.3.3 Jungle of Assessment Methods of Social Value and Social Impact within the Literature

Extending the complications of terminology, assessment methods, and focus within these methods is a tangled jungle of approaches. Scholars describe the difficult, if not impossible, task of collecting and comparing data on social value creation from unrelated heterogeneous activities [26]. Kroeger et al. in [26] go so far as to conclude that current approaches do not reflect actual social value creation nor offer the value of comparison.

Table 2.2 lists the assessment method focus in the literature studies. Be reminded that all literature studies selected assessed social value and social impact at a societal level. Six of the works of literature reviewed focused on the UN Sustainable Development Goals (SDGs), including sustainable impact assessment (SIA) approaches for measuring the economic, environmental, and social impacts of proposed policies, projects, and programs.

Four of the documents reviewed used a finance-based approach such as global reporting initiative (GRI) standards and/or Environmental, Social, and Corporate Governance (ESG) when sustainable investment is involved. Subjective Wellbeing (SWB), the Social-ecological system (SES) framework, and the Capabilities Approach introduce a unique methodology and focus. Hybrid combinations of assessment methods customized for the unique perspectives of each stakeholder are common.

Social Value assessments are often, but not always, linked to the terms corporate sustainability reporting (CSR) or simply sustainability. The United Nations (UN) provides the most predominant sustainability guidance for measurement and monitoring. According to Business Call to Action and GRI (2016), the UN guides businesses for social value measurements with the following stated purpose: "To offer governments who are engaging with the private sector how business tools, impact measurement, and sustainability reporting, can be used to measure, monitor and accelerate the business contribution to the Global Goals" (p. 1).

Table 2.2 Assessment method focus in literature studies

Citation	Stake-holder view		SDG	Assessment method focus			Other/comment
	Business	Institutions	Sustainable development goals (SDGs)	SIA Based (EIA, SA, PSIA)	Finance based approaches (Social value is measured in monetary terms)	Subjective well-being (SWB) including life satisfaction (LS)	
Business call to action [6]	X	X	X	X	X		SIA for assessment GRI for business reporting
Colantonio [8]		X	X	X			
Domínguez et al. [12]	X	X	X	X			
Goedkoop et al. [19]	X		X	X			
Kato et al. [24]	X	X					Capabilities approach
Kroeger et al. [26]	X					X	
Maas et al. [28]	X	X					Plus a classification system includes additional six characteristics

Reference							Social-ecological system (SES) framework / Notes
McGinnis et al. [29]		X					Social-ecological system (SES) framework
Mulgan [30]	X				X		
Polonsky et al. [33]	X				X		Four different approaches: operating efficiency; achievement of organizational objectives; return on investment; and social outcomes.
Porter et al. [34]	X				X		Business results combined with social results
Potma [35]	X					X	
Tuan [42]	X	X			X		
Vanclay et al. [45]			X	X			
World business council [46]	X		X	X			

The remaining studies layered bespoke assessments targeting the unique goals of the party conducting the assessment. However, hundreds of competing assessments, metrics, and tools for measuring social value are available [30]. Some are developed for non-profits, others for corporations, and others for governments. Methods are often tailored to the organization conducting the assessment's requirements based on the impacts they seek to measure.

Maas and Liket present 30 social impact measurement methods on Social Impact Measurement: classification of methods as consolidated in one of the studies of this landscape theory review [28]. They note that existing assessment methods do not take a common understanding based on what is measured, why it is measured, for whom it is measured, and how to measure it [28].

In Maas et al. [28] study on assessment characteristics, almost all methods had multiple purposes. Over half (17/30) are appropriate for screening. Over half (18/39) are suitable for monitoring. The dominant purpose of the assessment was reporting (24/30). The most dominant purpose categorized was evaluation (25/30). 9 of the 30 measures could be used for all purposes [28]. Moreover, only six of the 30 methods analyze the social impact from a micro, meso, and macro perspective.

Maas and Liket's research indicates that only 8 of 30 measure social impact. These eight methods all have a macro (society) perspective. Four additional methods partially measure impact; however, this is less than half of the 30 assessments (12/30). Most of the methods orient their approach toward inputs rather than outputs.

Maas and Liket sought to unravel the and add structure to account for the differences across assessment methods [28]. They break down their findings that assessment method characteristics differ from a stakeholder perspective and across dimensions of purpose (screening, monitoring, reporting, or evaluation), time frame orientation (prospective, ongoing, or retrospective), length of time frame (short-term or long-term), perspective (individual/micro, corporate/meso or society/macro), and approach (process methods, impact methods or monetarization) [28].

These characteristics and type classifications are useful for conceptual model representing social sustainability interconnections in an information systems net-work. All dimensions may not be necessary for every query or project; however, all types of dimensions hold relevance for labeling information into data sets that a computer with algorithms can understand. However, real-world problems are not easily described with the precision that social impact approaches (and computer algorithms) insist upon [30]. There is an inherent difficulty in giving mathematical expression to the complexity of human behavior due to limitations and the costs of collecting and analyzing data [30].

Two studies with a unique assessment focus researched are the Subjective well-being framework (SWB) and the Social-ecological system (SES) framework.

The SWB distinguishes between social need value creation and non-social need value creation, which offers flexibility for commercial, cultural, or other types of value creation [35].

The Subjective Well-being Framework for comparing social value creation includes subjective satisfaction constructs that offer uniform units of measure applied to social initiatives and businesses that wish to compare social performance with competitors [35]. The SWB framework considers the individual, community, and broader society they live within. Life satisfaction (LS) denotes living standards within a group, community, state, or nation [26].

It also applies to business as the framework allows for the segmentation of people served or customers while also affording groupings for disadvantaged individuals or groups (Domains) whom business activities may have negatively impacted.

The definition of life satisfaction (LS) is the deviation of an individual's achieved level of need from the aspired level of need [26] (see p. 520). LS is measured at the individual level and aggregated to a group or group of group's domains. LS is specifically personal, revealing a person and group's needs. DS provides a direct connection between the Impacts of change activities on the LS of the treatment group of actors involved in the change activity.

The SWB framework is a mathematics-based model with six measurement steps to determine SVC within specific time frames, which aligns well with the automation of social value's systemic integration. SWB offers an empirical design to capture social value, including indexing a control group for data to quantify and capture what would have been achieved without the initiative. The assessment model considers social domain needs across seven criteria: health, education, financial situation, community integration, housing, equality, and safety.

The second outlier is the social-ecological system (SES) framework with an institutional focus from academia. It emphasizes the need for researchers from diverse disciplinary backgrounds, working within different resource sectors from disparate geographic areas, to consolidate and share a common vocabulary [29]. This shared vocabulary is crucial for constructing and testing alternative theories and models to understand the influences on processes and outcomes.

The framework consists of ten subsystem variables that affect the probability of self-organization activities toward social and ecological sustainability. It was developed in collaboration with scholars and members of the resilience alliance in 2007. It has since been improved and expanded to incorporate policy and governance systems into the modeling [29]. The updated framework supports scholars and policymakers in gathering knowledge from empirical research and integrating analytical, diagnostic, and prescriptive capabilities [29].

The system comprises various components such as Resource Systems, Resource Units, Governance Systems, and actors. It acknowledges that while it models within a system that measures as a whole, external influences can affect components within the SES. For example, processes or operations vary in scale within the SES focus.

This reinforces and builds upon the Building Block I requirement that not only terminology and definitions but also categories, labeling, classifications, characteristics, and types are across assessment methods, and stakeholder groups are needed.

The information system network must connect the information to support decision-making and cross-disciplinary data integration for socially relevant data. For cost efficiency and improved data consistency, automation of standard assessment methods for data gathering is considered for conceptual modeling.

Additionally, policymakers will need guidance with analytical, diagnostic, and prescriptive insights from expert groups in academia to build the models that direct the technical systems to maximize social benefits and avoid negative impacts.

A use case for conceptual modeling would be to ensure that the model allows policymakers to receive guidance with analytical, diagnostic, and prescriptive insights from expert groups in academia.

The conceptual model should include a robust prototyping and testing arena within the network for debate, discussion, and cooperation in the development of automation of social value quantification data capture with the capacity for social sustainability risks monitoring, management, and reporting.

2.3.4 Perspective Matters When Measuring Social Value and Social Impact

Assessment has to do with an evaluation. Perception considers our understanding of the relationship between factors as a whole. How the assessor views the problem requires a perspective. This subjective point of view can hold notable significance when measuring social value and social impact on social sustainability. Stakeholders each assess for a reason and from their perspective.

The literature was further explored for similarities and differences because each stakeholder group measures social value and social impact. Table 2.3 lists and compares the reasons documented for measuring social value and social impact across the literature studies.

These studies show that parties measuring social value often have several reasons for the assessment, influencing their perspective. All studies noted a business relevance to social value/social impact measurement. Both stakeholder groups' top reasons for assessment were (1) decision-making, (2) cross-disciplinary data integration, and (3) risk reporting. Literature from institutional stakeholders more frequently noted a reason for social value measurement to harmonize definitions. In comparison, business was further interested in the comparability of data against competitors.

Institutions are measuring social value with more focus on accountability and monitoring social performance [6, 38, 40, 41]. According to [12], research indicates that "Growth in academic production and professional application of Social Impact Assessment (SIA) have indicated increasing interest in risk and impact analysis on the part of the scientific and business communities" (p. 1).

Measuring social value and impact is commonplace across the social sector. Infrastructure, mega-projects, and associated business/institutional activities create benefits and harmful effects. Large-scale projects have a differential distribution of costs and benefits across various communities. Measuring social value may not capture all variables but is expected to support sound management. According to

Vanclay et al. [45], "to ensure that the benefits of projects are maximized, and the negative impacts are avoided or minimized" (p. 1).

Through the SDGs social value, measurements are primarily focused on factors for stability and capacity development at a enduring level but not necessarily thriving. Business is also interested in aspects related to enrichment value such as luxury, comfort, life satisfaction, and experience beyond that which provides stability. These are also quantifiable social value criteria, although not a part of the sustainable development goals.

What are the reasons "why" business assesses social value? According to Mulgan [30], "The demand for measuring social value comes from all sides: funders who want to direct their money to the most effective projects, policymakers, and government officials have to account for their spending decisions, and social organizations need to demonstrate their impact to funding bodies and investors, partners and beneficiaries" (p. 1).

Michael Porter anticipates an increased business focus on solving global social problems to achieve a competitive advantage [34]. According to Porter et al. [34], "Companies create shared value by developing profitable business strategies that deliver tangible social benefits. This thinking is creating major new opportunities for profit and competitive advantage simultaneously as it benefits society by unleashing the power of business to help solve fundamental global problems" (p. 1).

Organizations seeking to alleviate social problems often depend on third-party support, including investors, foundations, corporations, and governmental institutions. These sources have high expectations and demands regarding transparency and accountability, including social impact [26, 33]. Such businesses measure social value for integrity, sound business decision-making, and secure funding. According to Goedkoop et al. [19], "measure social impacts of products and services, in a way that recognizes its high quality, credibility and business viability" (p. 18).

Businesses, more often, translate social values into economic values. According to Kato et al. [24], "Although tools exist to measure social value, they tend to focus on converting non-monetary costs and benefits into monetary terms to demonstrate the cost-effectiveness of operations" (p. 1).

Businesses' profit orientation creates motivation and incentives to measure the value created within their target market, seeking opportunities to generate economic value. Business aims to exploit opportunities for competitive advantage. There is less incentive to track the adverse outcomes of communities outside this lens.

This subjective perspective can lend itself to bias and can hold notable significance when measuring social sustainability across society. To meet the systemic stakeholder demand above, an information system network for social sustainability must connect the information in an objective and consistent manner that supports decision-making and cross-disciplinary socially relevant data integration.

Further exploration investigated the known obstacles in assessing and measuring social value and impact.

Table 2.4 lists and compares the challenges for assessing social value and social impact as documented in the literature study.

Table 2.3 Reasons for measuring social value and social impact in literature studies

Citation	Stake-holder view		SDG	Reason for measuring				Focus		Comparability (competitive)	Product focus	Risk reporting (accountability)	Governance and regulation
	Business	Institutions	Sustainable development Goals (SDGs)	Data integration (cross discipline)	Definition harmonization	Decision making	Strategic competitive advantage	Business to solve social problems	Required for investment/ funding				
Business call to action [6]	X	X	Yes	X	X	X		X		X		X	X
Domínguez-Gómez [12]	X	X	Yes	X		X						X	
Kato et al. [24]	X	X		X		X							
Maas et al. [28]	X	X		X	X	X			X	X		X	
Mulgan [30]	X	X				X	X		X				X
Tuan [42]	X	X				X			X	X		X	
Vanclay et al. [45]	X	X	Yes	X	X	X		X		X		X	X
Goedkoop et al. [19]	X		Yes		X	X	X	X		X	X		

Kroeger et al. [26]	X				X	X		X				
Polonsky et al. [33]	X				X		X	X		X		X
Porter et al. [34]	X				X	X	X	X	X			
Potma [35]	X			X	X		X	X	X		X	
World business council for sustainable development [46]	X		Yes	X	X					X	X	X
Colantonio [8]		X	Yes	X	X							
McGinnis et al. [29]		X		X	X		X				X	X

Table 2.4 Challenges measuring social value and social impact across literature studies

Citation	Stake-holder view		SDG	Challenges for measuring								
	Business	Institutions	Sustainable development Goals (SDGs)	Existing meassurements are inadequate	Lack of common language	Diverging approaches	Short-term vs long-term impact	Positive vs negative impact not assessed	Methodological limitations	Manipluation and selective scoping	Lack of sustainability integration	Poor quality of lack of data and comparability
Business call to action [6]	X	X	Yes	X	X	X			X		X	X
Domínguez et al. [12]	X	X	Yes	X	X						X	
Kato et al. [24]	X	X		X	X	X	X		X			X
Maas et al. [28]	X	X		X	X		X	X				
Mulgan [30]	X	X			X				X			X
Tuan [42]	X	X		X	X	X			X	X		
Vanclay et al. [45]	X	X	Yes	X	X	X		X	X			X
Goedkoop et al. [19]	X		Yes		X	X						X
Kroeger et al. [26]	X				X	X			X			

Business call to action [6]	X	X	Yes	X	X	X		X		X	X
Polonsky et al. [33]				X	X						
Porter et al. [34]				X	X			X			X
Potma [35]			Yes	X	X	X		X	X		X
World business council for sustainable development [46]							X				
Colantonio [8]	X	X	Yes	X	X	X				X	
McGinnis et al. [29]	X	X		X	X	X		X	X		X

Similarities related to the challenges of measuring social value were also many across both stakeholder groups. Both business and institutional literature noted (1) Lack of common language, (2) inadequacy of existing measurements, (3) methodological limitations including diverging approaches, and (4) poor quality of data and lack of data comparability as the most noted challenges.

Institutions are calling for integrated approaches with business. Sustainability reporting and impact measurement are two practices companies use to improve performance, account for the impact, and publicly communicate sustainability data [6]. Institutions are particularly interested in socio-environmental relationships in multi-national organizations and large-scale development projects [12].

The information system network must connect the information in an objective and consistent manner, supporting decision-making and cross-disciplinary data integration for socially relevant data.

The model should have an objective system design with social sustainability as an overarching purpose.

A use case for conceptual modeling review is to verify that businesses can build strategies that target social problems with data on social goals and capture and quantify social value distributed across the information system network.

A use case for conceptual model design is that the model should be designed to influence the maximization of social benefits and the avoidance of negative impacts.

A use case for the conceptual building is that all stakeholders can use the data, potentially in real time, for decision-making; the information system network should be open and share information, including business comparability of data against competitors.

2.4 Building Block 1: Results for Application into Conceptual Modeling

2.4.1 Relationship Mapping

Mapping patterns and interrelationships of stakeholders is a visual tool for simplifying and exploring the complexity of how businesses and institutions are connected within the literature. Investigating patterns and influences allows a view into how each party assessing social value/impact perceives their purpose and the interdependencies with other stakeholders.

2.4.1.1 Relationships and Interdependencies

The stakeholder relationship diagram below in Fig. 2.1 summarizes the general expectations uncovered in the existing literature on the business and institutional

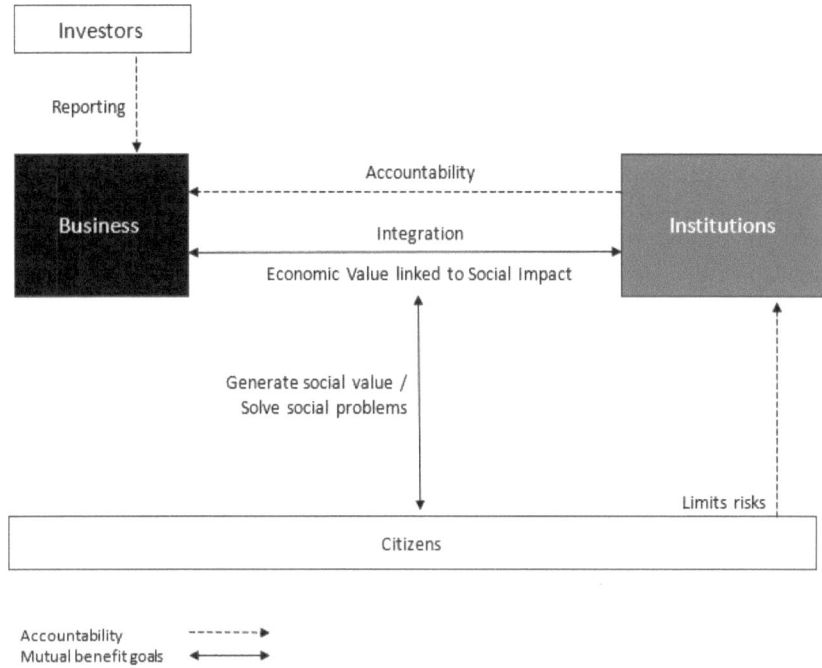

Fig. 2.1 Stakeholder relationship diagram

stakeholders' relationship goals and accountability structures. It strips down the assumptions to a minimalistic form.

Something to note is that amid the dynamic complexity of global societies, the theme of social sustainability is moving away from the Western concepts of equality and fairness toward limiting risks when calculating social value. According to Colantonio [8], "Sustainability indicators are increasingly focused on measuring emerging themes rather than improving the assessment of more traditional concepts such as equity and fairness. Indeed, the latter continues to be measured mainly in terms of income distribution and other monetary variables, hampering meaningful progress in social sustainability assessment" (p. 13).

2.4.1.2 Predominant Assessment Method Mapping of Social Value and Social Impact Across the Literature

Across this research's social value assessment literature, there were overlapping and additional assessment models and frameworks. The most predominant usage of standards, principles, assessment models, or frameworks based upon the interaction or relationship position are:

- Base of the Pyramid Impact Assessment Framework (Bop IAF), the BoP Impact Assessment Framework aims to understand who at the base of the pyramid is impacted by BoP ventures and how they are affected. The framework is developed to evaluate and articulate impacts, guide strategy, and enable better investment decisions [28].
- Basic Efficiency Resource (BER) is a simple framework for measuring the performance of complex multi-component programs, campaigns, or activities.
- Capabilities Approach (CA) The capabilities approach is an explicitly normative framework with a social justice orientation at its core. As a tool, it allows the assessment of individual well-being and social institutions, policies, and contexts that may influence individual well-being [24].
- Environmental, social, and governance (ESG) principles are standards for socially conscious investors to screen potential investments and monitor operations.
- Global Reporting Initiative (GRI) standards offer a common global language for organizations to report sustainability impacts.
- International Integrated Reporting Council (IIRC), creating a globally accepted framework for a process that results in organization statements regarding value creation over time.
- Participatory Impact Assessment (PIA) is a process carried out by communities in partnership with NGO professional practitioners to evaluate the impacts of development interventions.
- Poverty and Social Impact Assessment (PSIA) is an analytical approach developed by the World Bank to assess the social impacts of policy reforms, emphasizing vulnerable populations.
- Shared Value acknowledges the responsibility of a company to recognize that societal needs, not just conventional economic needs, define markets and that the purpose of the corporation must be defined as creating shared value, not just profit, so society benefits as well as the company [34]. This view also acknowledges that social harms frequently create costs for firms in the form of social risks and, therefore, need to be carefully monitored and managed [45].
- Social Accounting is evaluating and accounting for the social and environmental measures of an organization's actions.
- Social-ecological System (SES) is a framework that enables researchers from various disciplinary backgrounds and sectors to construct and test theories and models in empirical settings to support policymakers in the complex and dynamic social and ecological policy theater.
- Social Cost-Benefit Analysis (SCBA) is a weighing-scale approach where all the benefits are weighed against the minuses or costs considering social issues.
- Social Return on Investment (SROI) measures change to people or organizations that experience or contribute to the change. It assigns monetary values to represent social, environmental, and economic outcomes.
- Subjective Wellbeing (SWB) measures people's cognitive and affective evaluations of their lives.

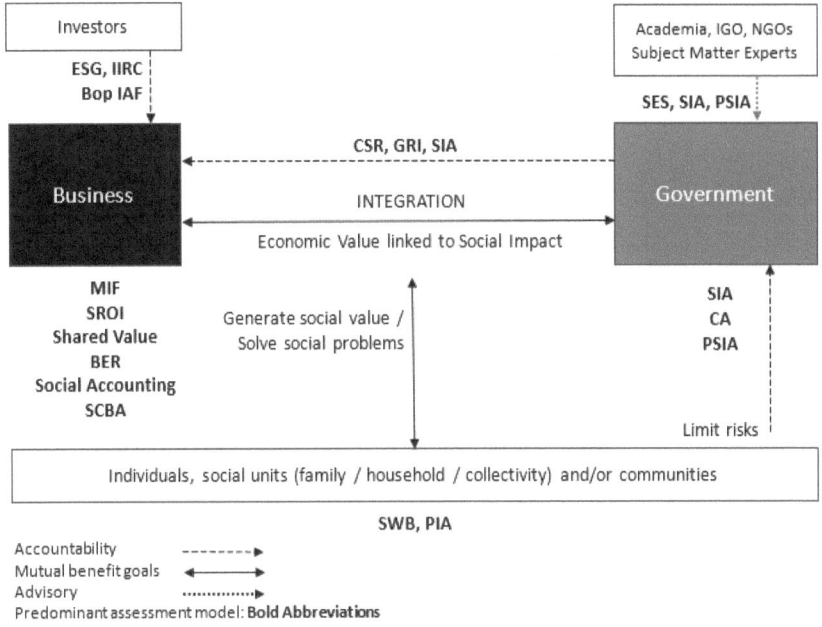

Fig. 2.2 Stakeholder relationships with predominant assessment methods

- Sustainability Impact Assessment (SIA) is a soft policy instrument and process for assessing policies, strategies, and programs' economic, social, and environmental effects for fully integrated sustainability in policy development.
- Measuring Impact Framework (MIF) helps corporations understand their contribution to society, assess and inform their operations, support investment decisions communicate with stakeholders.

Figure 2.2 maps these predominant assessment methods for systemic measurement of social value and social impact based upon which stakeholder relationship diagram.

Several global organizations are competing to establish social sustainability standards for investment in response to an urgent demand to improve consistency and comparability in sustainability reporting.

The International Sustainability Standards Board (ISSB) was formed by the International Financial Reporting Standards (IFRS) Foundation to replace the patchwork of voluntary guidance and establish a single set of global sustainability reporting norms for business [23]. Their effort focuses on investors' public policymakers, auditing firms, central banks, and service providers.

Furthermore, the Sustainability Accounting Standards Board (SASB) seeks to be a dynamic ecosystem of organizations, including investors, companies, policymakers, regulators, NGOs, and civil society, developing disclosure standards and frameworks for comparable, consistent, and reliable ESG information.

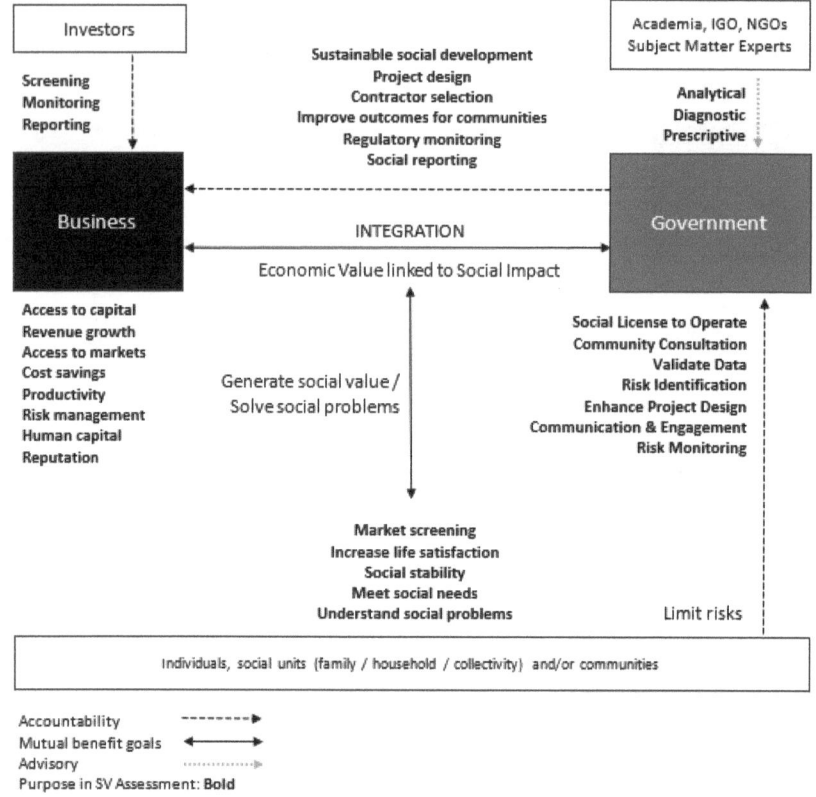

Fig. 2.3 Purpose of social value/social impact assessment by stakeholders

2.4.1.3   Purpose of Social Value and Social Impact Assessment Within the Literature Mapping

According to the 15 studies applied in this research, Fig. 2.3 diagrams a systemic perspective of the different purposes for assessing social value.

The issue of economics is front and center for businesses. Sustainability for businesses includes ensuring economic viability. Businesses must balance the economic and social impact goals and seek to monetize social value to speak the same language managers understand. Quantifying systemic data and processes for measuring social value needs to include a financial element of conversion when considering negative and positive impacts.

2.4.2 Setting Primary Definitions for the Conceptual Model

Analyzing the literature review with a new perspective provides a new lens for harmonization. Social value and social impact definitions are fluid and adapt over time to begin to take a form [1]. For computers and algorithm development for societal automation leveraging collective intelligence, consistency and clarity of primary terminology is required to ensure data quality and comparability.

The International Association for Impact Assessment (IAIA) was organized in 1980 and is the leading global network on best practices for social impact assessment (SIA). It brings together worldwide researchers and practitioners of impact assessment information regarding policies, programs, plans, and projects. SIA is applied to measure the social impact of megaproject development such as dams, mines, railways, large-scale infrastructure, and agriculture projects.

The IAIA definition of social impact has changed notably over the years. In the latest guidance, the IAIA breaks the terms into even more detail with various definitions, many of which relate to the time the assessment is conducted [45].

2.4.2.1 General Terms

- **Social impact:** something experienced or felt, in a perceptual or corporeal sense at the level of an individual, social unit (family/household/collectivity), or community/society [45].
- **Values:** abstract and often subconscious assumptions held by individuals about what is right and/or important in their lives. Typically, they are organized into a value system. Values and value systems can vary substantially between cultural groups [45].

Maas and Liket conducted a definition comparison in their review of the term social impact across thirty social impact measurement methods [28]. This research further notes the definition of social impact developed by Clark et al. 2004 "by the impact, we mean the portion of the total outcome that happened as a result of the activity of an organization, above and beyond that what would have happened anyway" is based upon the Impact Value Chain [7].

A baseline, a quantification of the current social valuation, is inferred in this context. A separation and evaluation of the positive and negative as well as short-term and long-term effects of an action or project on society are essential for assessing social impact [6, 18, 28, 30, 35, 45].

Aligning closely with the International Association for Impact Assessment's (IAIA) guidance but including Maas's input definitions comparison review, a clear separation of social value and social impact is with modifications in bold. A form of measurement is required; therefore, the specification of "quantification" is added for clarity.

The following applications of **Key Definitions** are applied in this research forward:

- **Social Value** is **the quantification** of the relative importance that is experienced or felt in a perceptual or corporeal sense at the level of an individual, social unit (family/household/collectivity), or community/society.
- **Social Impact: The portion of the total outcome** that is experienced or felt, in a perceptual or corporeal sense at the level of an individual, social unit (family/household/collectivity) or community/society **due to an event, project or change above and beyond what would have happened anyway.**

2.4.2.2   Screening Terms

- **Potential impact:** is an impact that is predicted rather than an actual impact that has already occurred [45].
- **Impact:** an economic, social, environmental, and other consequence that can be reasonably foreseen and measured in advance if a proposed action is implemented [45].
- **Impact equity:** the notion that the impacts in society or a project should be shared equitably, at least that there should be consideration given to the fair distribution of negative and positive impacts. For example, the adjustment of flight paths for an airport to share the noise burden rather than the same people having all the noise [45].
- **Perceived impact:** something that is believed to be a potential impact rather than something established as an actual impact. Note that perceived impacts affect how people feel about the project and how they feel and generally behave; thus, perception is the reality for them [45].

2.4.2.3   Monitoring Terms

- **Social change process:** an identifiable process of change in project-affected communities that are created, initiated, enabled, facilitated, or exacerbated by a planned intervention. The social change process is not in itself a social impact but may or may not result in the experience of social impacts depending on the local context. For example, migration and resettlement are social change processes that may or may not result in social impacts [45].
- **Direct impact:** is an impact that occurs as a direct result of the planned intervention. It may also be called primary impact or first-order impact. In SIA, it refers to social changes and social impacts caused directly by the project itself, such as the annoyance to people of noise generated by machinery associated with the project [45].
- **Actual impact:** refers to the social impacts actually being experienced by communities rather than to the impacts predicted [45].

2.4.2.4   Reporting Terms

- **Social footprint:** a concept that attempts to be the social equivalent of 'ecological footprint;' thus, a metaphor refers to the extent of social harm created by a project or product. It is a concept not favored by social scientists and thus not used in SIA, but some physical scientists promote it alongside the ecological footprint [45].
- **Shared value:** is a way of thinking that the purpose of the organization must be defined as creating shared value, not just profit, so both society and the organization benefit. This view also acknowledges that social harms frequently create costs in the form of social risks and, therefore, need to be carefully managed [45].
- **Indirect impact:** is an impact that occurs as a result of another change caused by a planned intervention. In SIA, an indirect effect might be caused by a physical change to the environment. For example, a mine might increase river turbidity, reducing the supply of fish and reducing the economic livelihoods of fishing-dependent villagers. These can also be secondary or higher-order effects [45].

2.4.3   Model Requirements for a Systemic Alignment of Social Sustainability Information in a Technological Network

The exploratory investigation with a synthesis approach provided insights to begin the thought trials through metaphor.

Stakeholders' expectations of interrelationships, purpose, challenges, and assessment methods provide insights into the interdependencies. These relationships and inter-dependencies represent technological requirements for developing a conceptual model for a social sustainability information system network.

The Building Block I Requirements are [1]:

1. The model should have an objective system design with social sustainability as an overarching purpose.
2. The information system network must connect the information to support decision-making and cross-disciplinary data integration for socially relevant data.
3. So that all stakeholders can use the data, potentially in real-time, for decision-making, the information and system network should be open and share information, including business comparability of data against competitors.
4. So that technology can integrate data with quality cross-disciplinary and consistency, clarification of primary terms is required. This includes categories, labeling, classifications, characteristics, and type across assessment methods.
5. The model should be designed with predictive and automation capabilities to maximize social benefits and avoid the negative impacts cost-effectively.
6. Business requires quantifying systemic social data and processes for measuring social value, including a financial conversion of positive and negative impacts.

7. The conceptual model should include a prototyping and testing arena within the network for societal debate, discussion, and expert collaboration.

References

1. Aebi C (2023) Business and societal requirements for an information system network for social sustainability. In: Research in sustainability, pp 281–295
2. Assefa G, Frostell B (2007) Social sustainability and social acceptance in technology assessment: a case study of energy technologies. Technol Soc 29(1):63–78
3. Blomquist W, DeLEON P (2011). The design and promise of the institutional analysis and development framework. Wiley, Hoboken
4. Brundtland GH (1987) Our common future—call for action. Environ Conservation 14(4):291–294
5. Bushouse BK (2011) Governance structures: using iad to understand variation in service delivery for club goods with information asymmetry. Policy Stud J 39(1):105–119
6. Business Call to Action and GRI (2016) Measuring impact: how business accelerates the sustainable development goals. United Nations Development Programme and GRI, Istanbul
7. Clark C, Rosenzweig W (2004) Double bottom line project report. University of California, Berkeley
8. Colantonio A (2009) Social sustainability: a review and critique of traditional versus emerging themes and assessment methods. Doctoral Dissertation, Loughborough University
9. Cunningham K, Ricks M (2004) Why measure: nonprofits use metrics to show that they are efficient but what if donors don't care. Stanford Soc Innov Rev 2:44–51
10. Deming WE (2018) The new economics for industry, government, education. MIT Press
11. Dillenburg S, Greene T, Erekson OH (2003) Approaching socially responsible investment with a comprehensive ratings scheme: total social impact. J Bus Ethics 43:167–177
12. Domínguez-Gómez JA (2016) Four conceptual issues to consider in integrating social and environmental factors in risk and impact assessments. Environ Impact Assess Rev 56:113–119
13. Drakakis-Smith D (1995) Third world cities: sustainable urban development 1. Urban Stud 32(4–5):659–677
14. Drucker P (1954). Management by objectives. Acad Manage Rev 6:225–230
15. Durkheim E, Spencer H, Comte A, Merton RK, Lazarsfeld P, Coleman J (1972) Political sociology. Selected Writings, pp 189–202
16. Emerson J, Wachowicz J, Chun S (2000) Social return on investment: exploring aspects of value creation in the nonprofit sector. In: The Box Set: Social Purpose Enterprises and Venture Philanthropy in the New Millennium, vol 2. REDF, Beaverton, pp 130–173
17. European Commission (2021) Corporate social responsibility & responsible business conduct. Retrieved from Internal Market, Industry, Entrepreneurship and SMEs https://ec.europa.eu/growth/industry/sustainability/corporate-social-responsibility-responsible-business-conducten.
18. FSG Impact (2013) Measuring shared value. Webinar, united states of america. Retrieved from https://youtu.be/RYuZcAVl0g
19. Goedkoop MJ, Indrane D, de Beer IM (2018) Handbook for product social impact assessment. Roundtable for Product Social Metrics Ed. Version, 4
20. Hardoy JE, Satterthwaite D (1991). Environmental problems of third world cities: a global issue ignored? Publ Admin Develop 11(4):341–361
21. Heikkila T, Schlager E, Davis MW (2011) The role of cross-scale institutional linkages in common pool resource management: assessing interstate river compacts. Policy Stud J 39(1):121–145

22. International Association for Impact Assessment (2022) Social impact assessment. Retrieved from IAIA https://www.iaia.org/wiki-details.php?ID=23
23. Jones H (2021) New global sustainability disclosure board draws heavyweight backing. Retrieved from Reuters: https://www.reuters.com/business/sustainable-business/former-ecb-head-trichet-advise-new-global-sustainability-disclosures-board-2021-06-07/
24. Kato S, Ashley SR, Weaver RL (2018) Insights for measuring social value: classification of measures related to the capabilities approach. VOLUNTAS Int J Volunt Nonprofit Org 29:558–573
25. Kitchenham B, Charters S (2007) Guidelines for performing systematic literature reviews in software engineering. https://legacyfileshare.elsevier.com/promismisc/525444systematicreviewsguide.pdf. Keele University; Durham University Joint Report, (EBSE-2007-01), Keele
26. Kroeger A, Weber C (2014). Developing a conceptual framework for comparing social value creation. Acad Manage Rev 39(4):513–540
27. Liu S (2017) Bioprocess engineering, 2nd ed. Elsevier, Amsterdam. https://doi.org/10.1016/B978-0-444-63783-3.09993-7
28. Maas K, Liket K (2011) Social impact measurement: classification of methods. In: Environmental management accounting and supply chain management, Springer, pp 171–202.
29. McGinnis MD, Ostrom E (2014) Social-ecological system framework: initial changes and continuing challenges. Ecol Soc 19(2):12pp
30. Mulgan G (2010) Measuring social value. Stanford Soc Innov Rev 8(3):38–43
31. Oakerson RJ, Parks RB (2011) The study of local public economies: multi-organizational, multi-level institutional analysis and development. Policy Stud J 39(1):147–167
32. Ostrom E (2009) A general framework for analyzing sustainability of social-ecological systems. Science 325(5939):419–422
33. Polonsky M, Grau SL (2011) Assessing the social impact of charitable organizations—four alternative approaches. Int J Nonprofit Volunt Sector Market 16(2):195–211
34. Porter ME, Hills G, Pfitzer M, Patscheke S, Hawkins E (2012) Measuring shared value: how to unlock value by linking business and social results. Foundation Strategy Group (FSG) (46910)
35. Potma L (2016) Social impact measurement methods. In: Important Indicators, Strengths, Weaknesses and Value Placed on Comparing Impact, University of Amsterdam, Amsterdam
36. Raworth K (2012) A safe and just space for humanity: can we live within the doughnut? Oxfam, Oxford
37. Rockström J, Steffen W, Noone K, Persson Å, Chapin FS, III, Lambin E, Lenton TM, Scheffer M, Folke C, Schellnhuber HJ, et al (2009). Planetary boundaries: exploring the safe operating space for humanity. Ecol Soc 14(2):32
38. Rowan M, Streather T (2011) Converting project risks to development opportunities through sia enhancement measures: a practitioner perspective. Impact Assessment Project Appraisal 29(3):217–230
39. Shaw J (2011) One truth: a revolution in corporate reporting. Retrieved March 6, 2011, from Harvard Magazine: https://www.harvardmagazine.com/2011/02/revolution-in-corporate-reporting
40. Smith SL, Pollnac RB, Colburn LL, Olson J (2011) Classification of coastal communities reporting commercial fish landings in the us northeast region: developing and testing a methodology. Fish Rev 73(2):41–61
41. Tajziehchi S, Karbassi A (2011) Problems and challenges facing developing countries in order to execute the social impact assessment of dams—a review. Eur J Sci Res 56:489–495
42. Tuan M (2008) Measuring and/or estimating social value creation: insights into eight integrated cost approaches. Gates Foundation Website
43. United Nations (2019) Sustainable development goals. Retrieved from About the Sustainable Development Goals: https://www.un.org/sustainabledevelopment/sustainable-development-goals/
44. Vanclay F (2003) International principles for social impact assessment. Impact Assess Project Appraisal 21(1):5–12

45. Vanclay F, Esteves AM, Aucamp I, Franks D (2015) Social impact assessment: guidance for assessing and managing the social impacts of projects. International Association for Impact Assessment, Fargo
46. World Business Council for Sustainable Development (2008) Measuring impact beyond the bottom line: why measuring impacts on society makes business sense. WBSCD, Conches-Geneva

Chapter 3
Metaphor-Driven Contextualization: Reordering Sustainability Information for Collective Intelligence

3.1 Metaphor Flux and Transformation

Chapter 2 dove into social value/social impact literature and applied an exploratory synthesis approach to map mapped Social Impact Information. This uncovered stakeholder patterns such as relationships, purpose, and challenges for measuring social value/social impact by business and institutional stakeholders. It further uncovered the socially constructed reality view of the need for social information across an information system network establishing foundational Model Requirements. Data extraction and synthesis continue throughout the iterative research process with constant comparison, focusing on the relationships between stakeholder groups and allowing a visual representation and early theoretical story to emerge.

In Building Block II, this research considers the patterns and delves deeper into relationships with consideration for the general requirements identified in Building Block I. Metaphorical reasoning considers the environment of change and dynamic social complexity to explore further a pragmatic technical solution to meet these requirements within the context of an ever changing society.

The patterns of **social sustainability** information and relationships of societal parties was investigated within the formal and informal processes and systems. This exploration includes scalable thinking technology which may enable an information network to facilitate decision-making focused on stable and thriving communities within a consideration of dynamic complexity and the paradox of Eternal Change. **Systems thinking** offers the analytic skills to understand the social systems, predict behaviors, and devise modifications to produce desired effects. **System dynamics** offers the analytical capabilities to get to the roots of complex societal activities to introduce predictive capabilities for policy design and feedback loops to monitor and continuously adapt to improve social outcomes. **Stigmergic Systems** can act as a scale-free architecture for self-organization processes. Stigmergic systems offer the opportunity for framing groups of software services based on proposed fields

C. Aebi, *Unifying Sustainability Information for Societal Automation*, SpringerBriefs in Business, https://doi.org/10.1007/978-3-031-83120-1_3

of activity to enable appropriate storage and operating solution stack configurations for the deployment of services.

Theorizing consists of disciplined imagination, thought trials, and elaborations, for which metaphor is a valuable tool to focus attention. Metaphorical reasoning builds an abstract and generalized sensory representation by creating an imaginary experiment for hypothetical scenarios [66]. Metaphorical reasoning engages in constant comparison to move from a sensory to a conceptual representation by providing the vocabulary to express and visualize the relationships in a provisional way of organizing the constructs [8]. Further, metaphors offer a recognizable bridge of analogy to direct attention to the subtle parallels while retaining an openness of meaning to fill in further the emerging theory's unique details [8, 55].

Technological advances are moving research toward modeling, concept development, and predictive thought. Metaphors are a vital part of this process, given that the subject matter in automating technology for social sustainability across society is complex. The use of metaphor circumvents the limits of methodological structures that favor validation rather than usefulness, avoiding trivial theories [55].

The metaphor applied for this research is sourced from the book Images of Organizations, initially written by Gareth Morgan in 1986 with an updated edition in 2007. Its audience includes management practitioners, educational institutions, and leaders of change initiatives. In this era of disruptive innovation and adaptive problem solving, Morgan offers understanding to support managers with approaches to complex situations with flexible thinking. The reader experiences organizations and management through the lens of metaphor by asserting that A is similar to B. Storytelling with patterns and similarities offers easy-to-grasp rich imagery. A metaphor is not to be taken as fact or complete. It is, by nature, a distortion to support the construction of an image.

The book categorizes organization into the following metaphors: (1) machines, (2) organisms, (3) brains, (4) cultures, (5) political systems, (6) psychic prisons, (7) flux and transformation, and (8) instruments of domination. The longest and most complex chapter describes the metaphor of flux and transformation [40].

The Images of Organization's metaphor "flux and transformation" lens/prism metaphor was selected due to the constant flux and transformation of social value alongside the interconnections and demand for social stability and continual change in social development. Morgan posed two complementary metaphors that were considered for this research.

Firstly, Organism, and secondly, Flux and Transformation. Both look at the interaction between "living" and the environment. Both also consider inputs and outputs to meet survival conditions where environmental changes present problems that trigger a reaction. There is general agreement between the two metaphors that changes in the external environment are primary influences on adaptation and change.

One key difference between the Organism and the Flux and Transformation metaphor is that the Organism has one life and dies. In Flux and Transformation, the "living organization" can transform and be reborn through self-re/production. For example, Dyson, a company initially well known as a company that produces vacuum cleaners, added on many complementary product lines and has transformed into an electric car company. A product line may live out a life cycle, but the organization, if consistently adapting to its environmental needs, it can continue to renew itself. As society is constantly renewing and redefining itself with no end, and we are at a stage of unprecedented change and transformation, the metaphor Flux and Transformation was selected to focus attention in this study. The metaphor "Flux and Transformation" draws comparative imagery of an organization's existence within the universe. Organizations experience unseen forces that link, implicate, and explicate orders as a never-ending process of flux and transformation.

Societal complexity defies comprehensive analysis [41]. The metaphor Flux and Transformation is broken into four processes in the logic of change to offer the imagery and perspective to observe and explain the tensions and paradoxes that shape them. Morgan gives examples of achieving quantum change holistically by investigating the nature of coherent self-organization for non-linear systemic change.

Flux and transformation is a centuries-old theme that remains relevant and alive today. Lao Tzu noted around 500 B.C in ancient China, "Life is a series of natural and spontaneous changes." Do not resist them; that only creates sorrow. Let reality be reality." The Greek philosopher Heraclitus remarked, "You cannot step twice into the same river, for other waters are continually flowing on, everything flows, and nothing abides; everything gives way, and nothing stays fixed" It is in changing those things that one finds repose." Contemporary theoretical physicist David Bohm proposes a theory for understanding the universe as flowing and whole. His investigation demonstrates that process, flux, and change are at the essence of our reality. There is an implicate order of potential not yet realized and an explicate order already manifested. Desires or ideas, big or small, create the expressed potential (implicate) that generates momentum to create or change a reality (explicate). Simultaneous forces of chaos (entropy) and order (negentropy) bind our solar system and the universe.

These tensions and patterns of unity and disorder are also observable within a business and social system. A constant state of flux applies ever-new pressures and momentum for transformation. In the resulting pocket of disorder, the incentive or focus of the effort reinstates the status quo or achieves a new state of order.

Morgan organizes the metaphor of Flux and Transformation into the following four processes in the "logic of change" [40]:

- Relationships between systems and their environments
- Chaos and complexity theory
- Cybernetics
- Change as the tensions between opposites

These four processes in the logic of change offer the imagery and perspective to observe and explain the tensions and paradoxes that shape these Flux and Transformation processes.

3.1.1 Relationships Between Systems and Their Environments

This logic of change begins with the term "autopoiesis." Humberto Maturana and Francisco Varela introduced this term in the 1973 book Autopoiesis and Cognition: The Realization of the Living. Autopoiesis is derived from the root words "Auto," which means self, same, or one in Greek, combined with "poiesis," which means creation or production. He saw a need to define what takes place in the dynamics of autonomy within living systems. When applying the term to an organization, "autopoiesis" refers to its ability to self-create and reproduce to maintain itself.

Maturana and Varela offer a new perspective, concluding that all living systems are organizationally closed, autonomous systems [32]. The Autopoiesis concept distinguishes the system and the environment, stating that the system is only open to viewing and making sense of the environment to gain new perspectives for understanding the existing processes requiring adaptation.

Living systems have the characteristics of autonomy, circularity, and self-reference [32]. These characteristics afford the ability to self-create and self-renew [32]. Living systems aim to reproduce themselves: their organization and identity are the essential products [32]. A system's interaction with its environment is simply a circular interaction process exclusively to facilitate self-production [32].

This description does not imply that the systems are isolated. The closure and autonomy are organizational in distinguishing the system as a "system," which may be within multiple additional layers of interacting and often inter-dependent "systems." The theory of autopoiesis acknowledges that living systems have "environments" but states that the relationship with their environment is internally determined. There are numerous options for chains of interaction between systems. However, there is no dependent pattern of causation. Due to self-determination, changes in company A do not force changes in company B, C, D, or E. However, managing within a complex system also implies that no one is in the control position comprehensively [40].

One of the most notable challenges for applying the autopoiesis theory is defining each "system" across society. *Society* is a system that seeks to self-create and self-renew. According to Morgan, each person within a society is a system under this definition. Individuals self-create and self-renew through several stages of life: childhood, adolescence, career, and maturity. This process may create numerous identities throughout a lifetime.

Further, any group of individuals is a system, for example, a family, a club, a team, or a company. These groupings are also autonomous, circularly self-referencing, and able to self-create and self-renew. The individual remains an autonomous system operating/living within various autonomous systems, operating and living within even larger autonomous systems of states, countries, and even the planet.

Systemic alignment of social data must consider the micro (individual) that interact together in groups or organizations (meso) to build and work within the order of the macro-system (society).

When considering research in systemic alignment for social value for the application within societal automation, it is vital to consider the structures that seek and adjust toward equilibrium. Computer-augmented intelligence should contribute to harmonious cooperation between interacting parties to encourage the synchronization of activities across the system. Data sets of social aspects can bring to light patterns for organizations to connect and align parties, processes, and practices with the interdependent systems that, in the end, sustain them.

One missing piece brought out in the metaphor relationships between systems and their environment is including the individual in social sustainability for societal automation as a fundamental micro system. A look into the literature and this brings the investigation into Human Computation.

3.1.1.1 Human Computation

Human computation theories are advancing capabilities for analyzing complex and dynamic systems, offering a new lens for systemic integration of social value for systemic integration modeling. Human computation modeling is a broad and emerging field. The Handbook of Human Computation is a collection of human computation (HC) research across diverse areas of inquiry and application [35]. HC "thinking systems" are capable of sophisticated problem-solving at a level of complexity beyond any person.

The organization of society consists of adaptive, complex, and dynamic systems. Social processes interact across numerous variables. Social models include adaptive agents influencing one another in response to the stimulus they receive. Simple and anticipated local interactions can create familiar but perplexing larger patterns in norms, communication, or collective action. The emerging social patterns may take hold or disappear as dramatically as they appear. Human computation models consider "actors" within social order and interaction. Macro sociological theories stipulate that network topology and social stratification align structural factors to achieve social outcomes [30]. HC offers the potential to guide local individual coherent behaviors toward desired collective global performance through rules and incentive structures that reinforce solutions that maximize collective benefit [35]. Engineered incentives theoretically balance the individual goals in synergy with the collective level goals to avoid and self-correct in the event of exploitation.

The foundation of a social system begins with the connection between individuals [10]. Individual humans are a unique, complex system connected to others across a social system. Each individual can be viewed as a fundamental processing element of a more extensive social system [35]. A human social organization is a hierarchy of individuals connected within a complex network of family, friends, professional colleagues, and associations.

Like any other organism, the system provides a stable organization with an adaptive capacity to optimize its ability to survive in changing environmental conditions [35]. To pursue stability, the system operates with established rules, within limitations, where flexible and adaptive responses occur. At both local and global scales, social systems undergo change and more significant phase transitions of social hierarchy [22].

Complex systems display non-random associations over multiple temporal and spatial scales [64]. The property of criticality characterizes scale-free systems; that is, they sit on the edge between entirely predictable (consistent) and completely unpredictable (chaotic) behavior. A social system pursues both stability and a high degree of flexibility for adaptation. Such scale-free systems have a typical architecture of network elements called nodes, linked by connections called edges that describe human social organization and behavior [65].

Marvin Minsky is an expert in Computer Science and Cognitive Psychology and co-founder of the Artificial Intelligence Laboratory at MIT. His work considers large-scale distributed thinking systems and their connection to create intelligence as an emergent property [36]. Minsky concluded that intelligence is not created from intellectual skills but from the patterns within which intellectual skills are connected [36]. Thinking consists of a string of processes, which Minsky refers to as Agents. The Minsky "Society of Mind" model explains the cognitive architecture of patterns of organization of agents, which result in the emergence of intelligent behavior.

Minsky imagines a system in which higher levels of intelligence are derived through complicated computational processes built from a set of crowd-sourced primitive unintelligent functions. Examples: pattern matching, difference identification, categorization, sorting, remembering, observing, questioning, simulating, predicting, optimizing, analogies, and acquiring new processes [35].

Minsky's groundbreaking work has recently resulted in a framework for representing knowledge [39]. The term frame represents a data structure for representing a situation. Associated with each frame are various pieces of information. Some of the information describes how to use the frame. Other information is about the frame output. Additional information is related to troubleshooting if the expected outcome is not confirmed.

These frames can be viewed as a network of nodes and relationships. There are fixed frames that represent what is always present in any situation. Then, additional levels of relationship terminals or "slots" are dynamic and are applied only in specific instances or data [39]. Each terminal has particular conditions or "markers" that require the terminal to be assigned to a person or an object with a specific type or sub-frame. Several terminals can be assigned for more complex conditions and relations. The linked collections of related frames are called a "frame-system" [39]. Transformations mirror the effects of actions between frames within a system. Value calculations can represent and account for these transformation changes. Information can be gathered, shared, and coordinated across frames through a system of terminals.

3.1.1.2 Self-Organizing Systems

Integration and inclusion of individual (micro scale) data into information that remains confidential yet is useful for organizations or governments requires a form of self-organization. Stigmergic systems have been researched across disciplines, with social complexity and coordination as a core theme. Scalable thinking digital technology's practical applications are increasing within cognitive systems such as supply logistics, computer networking, automated resource allocation, economies, browser search recommendations, and autonomous vehicle traffic navigation algorithms.

Stigmergy is a class of mechanisms that intermediate animal interactions [23, 63]. Pierre-Paul Grassé introduced this concept in 1959. As a zoologist, he sought to understand and explain organization mechanisms in insect societies. Individuals work independently and within coordinated collective activities at the social level of biological organization. Stigmergy was initially investigated in termite colonies and other animals, such as bird flocking patterns and fish schooling behaviors.

The mechanism of insect stigmergy is fundamentally simple: individuals leave modifications to the material structures or leave traces, in the form of pheromones, in their environment, which gives feedback to themselves and other insects. In part, the colony records its activity in the physical environment and uses it to organize collective behavior interactions dynamically [63].

Although software architecture for a system of systems is not within the scope of this research, consideration for a desirable outcome with long-term viability is expected when building a conceptual model for social sustainability.

Stigmergy can be used as a form of indirect interaction for systemic self-organization. It offers the opportunity to frame groups of software services based on the proposed conceptual model. Stigmergic information systems must be designed to identify key fields of activity (actor base, architecture, software ecosystem) influencing them [42]. This enables appropriate storage and operating stack configurations for service deployment.

A stigmergic platform acts as a facilitator for the self-organization process. Digital technologies manage the large scale of relatively simple agents to collectively address problems with robustness and adaptability to environmental variations, which would be too complex for a single agent, referred to under the discipline of swarm intelligence [17]. Swarm intelligence refers to group problem-solving ability emerging from the interactions of simple information-processing units [26]. Two types of swarm intelligence are seen in stigmergy-swarm formation results in the creation of groups, and synchronized formations result in a collective action or behavior.

In a stigmergic cycle, individuals indirectly interacting and directly communicating with their environment, which stimulates a response by others. Stigmergy does not give detailed instructions on the coordination or mechanisms for the response. The meta-process model demonstrates a minimum-spec indirect interaction between actors and the environment to proactively influence subsequent actions among other actors, leading to emergent behavior within a domain [42].

To achieve effective self-organization in a digitally supported stigmergic social system, it is crucial to focus on minimal specification within the chaos theory framework. Essential elements of this system include a software platform that enables users to create, modify, and share information. This platform functions as a regulatory system, facilitating user contributions' aggregation, stimulation, and distribution within the stigmergic information network [42]. Stigmergic information system architecture and processes include a software Ecosystem of symbiotic or dependent services and platforms; and Support components of the architecture [42].

In the human computation paradigm, the computer relies on humans to perform processing tasks [62]. Within the stigmergic information system, human actors may aggregate, contextualize, and select a structurally limited yet quantitatively unlimited amount of information [42]. The actor base is all humans consuming or producing within the stigmergic information system concept. Each human retains autonomy and specifies a relationship connection to a group, activity, or role. A contributor expects to own their data unless explicitly stated otherwise, having the control of informational self-determination [42].

The stigmergic information system relies on a functional architecture as an information network with control services. The information network captures homogeneous and heterogeneous data, depending on the task. Machine learning algorithms analyze, match, and offer suggested connections for human selection. The network is flexible for growth with new data sets, and organization intelligence increases through patterns of interconnectivity.

Control services manage workflows and information exchange between platforms, actors, and ecosystem services. According to Musil et al. [42], in the paper titled A First Account on Stigmergic Information Systems and Their Impact on Platform Development, there are four types of control services (p. 3):

1. Assimilation services: assist in the linking of external content into the information network, e.g., sharing buttons, bookmarks, etc. [42].
2. Access control services: specify which information is permitted and grant restricted access to external dependent-symbiotic services within the ecosystem through dedicated APIs, e.g., payment services, collaboration tools, etc. [42].
3. Meta-heuristic services are responsible for adjusting the stigmergy flow and stimulating the actor base directly to contribute content, e.g., reviews, rates, input, etc. Search techniques that iteratively seek a matching solution across a set of multiple complicated or disconnected possibilities, e.g., task recommendation, service alternatives, profile recommendations, etc. [42].
4. Mining and analyzing services: assess stored information and data within the network and provide knowledge regarding actor contributions. Collections of open-source software utilities may be stacked to facilitate parallel processing of large data sets, e.g., a scalable distributed file system that can evaluate worker quality such as efficacy, efficiency, and accuracy and target critical tasks to store across multiple machines without prior organization [42].

The software ecosystem is an environment of cooperating services interacting with the actor base. It includes dependent-symbiotic services and platforms dependent upon accessing the network platform and providing functionality that enriches the host platform's utility for the actor base.

In the stigmergic information systems there are various participants. Offering the most sizeable contribution is the actor base, which effectively act for themselves as producers and consumers. Ecosystem-dependent symbiotic services of the platform also exchange high volumes of information and interconnectivity acting as an additional participant category. Moderate inbound traffic is further expected from participating collaborating control services.

The complex behavior of communities can be facilitated through software architectures for self-organization through simple interaction processes. This research investigates which social sustainability information is required by which societal parties within formal and informal processes, systems, and relationships. It explores whether scale-free distributed technologies, theories, and models may enable an information network that considers the dilemma of dynamic societal complexity within the paradox of eternal change.

A stigmergic information system network offers the potential for social self-organization within the planetary boundaries, monitoring, and risk reporting with criteria that align with the minimum set of social boundary conditions to which all parties are subject.

3.1.2 Chaos and Complexity Theory

Autopoiesis considered the organization in relation to the environment. Next, we will examine the same interconnection, focusing on "evolving patterns."

Chaos theories and self-organization investigate complex non-linear systems such as societal and ecological systems. These systems consist of multiple interacting systems with characteristics of both order and chaos. Due to internal complexity, random disturbances can create unpredictable reverberations throughout the system. Interestingly, despite the unpredictability, coherent order always emerges from randomness and surface chaos [27].

Chaos theorists have noted that system behaviors fall under attractor patterns or influences. The entire system can flip between patterns [29, 56]. Unpredictable events stimulate patterns of behavior that create coherent forms repeatedly. Complex systems define the context in which detailed system behaviors unfold. Attractors can pull a system into equilibrium states (or near equilibrium), such as negative feedback loops that counteract destabilizing fluctuations.

Chaos theorists investigate when systems get pushed toward the edge of chaos. Then, the system's energy self-organizes through unpredictable leaps into different system states. If the new set of influences gains a critical mass of attraction, the energies pull to a new configuration. Bifurcation points are associated with attractors and exist as latent potentials in complex non-linear systems. They signal the potential for self-organization and evolution in a new form.

It is important to note that no one can be in control of the change by acting on the insights of chaos and complexity theory. Shoring and nurturing key elements and context to set an attractor is possible. Nevertheless, the form will be its own. There is no way to exert unilateral control over a complex system.

Mathematics from system dynamics can measure bifurcation values. As values approach bifurcation parameter/s in a previously stable system, these bifurcation points become disruptive attractors, and the system may flip into a new configuration. If a dominant attractor dissipates, possible changes may dissolve, and the system reverts to a pattern from the former state. The path of the system is entirely unpredictable because even a tiny change can unfold to create significant effects.

Considering emergence as a natural state, Morgan suggests developing confidence in self-organization. Technology offers a new view of society as a system of individual actors that can collaborate, matching needs to support spontaneous self-organization across broader pools of resources for synergistic transformation.

Chaos theory suggests the following thoughts for managing complex change:

- Rethink organization, specifically hierarchy, and control
- Assimilate the art of managing in continually changing contexts
- Build expertise in how to use minor changes for significant impacts
- Continuous transformation and emergent order are natural
- Consider new metaphors that aid in the self-organization process

From chaos and complexity, we learn the need to establish a fundamentally stable yet adaptive and flexible organization that allows room for continual change, allowing for a constantly evolving order that cannot be planned or predetermined. There is no grand design. Patterns will emerge organically without imposition.

3.1.2.1 The Complex Controller Structure

When considering the future option of computer-based decision-making for humans, Helmut Nechansky has diagrammed the process's epistemological aspects that determine individual decisions and behavior, titled the Complex Controller Structure. The complex controller structure models aspects of human reasoning with external looping for modeling decisions and internal looping for goal-value decisions [44].

In this model, the senses provide input data from the environment relevant to a goal value. The sensory input is used two-fold. Externally, it is used to scan and take goal-oriented action continuously. Internally, it also involves learning and occasionally modifying the goal values. Humans define what they want to achieve through goal-value setting at three levels:

1. Existential goal-values
2. Long-term goal-values
3. Short-term goal-values

A single decision or general hierarchy of priorities of models can be defined for computer-supported decision-making.

Machine learning models are applied to evolve from chains of historical sensor data identifying previously sequenced observable patterns, previous modeling decision overrides, and goal-value alignment for interrelated predictive sequencing.

Such models are trained for continual adaptation through confirmed and repeated observations of feedback loops from goal-value outcomes and contradictions to predicted goal states.

Advanced models integrate existential goal values within a hierarchy to offer predictions that individuals can use as an internal feedback loop to set long-term or short-term goal values.

The controller model can also be applied to the human interaction of two individuals. The process repeats with internal goal-value learning loops and external decision-action loops. This process either confirms that both parties share the same individual goal values or leads to the development of shared goal values, e.g., a parenting decision between father and mother or a business transaction deciding upon a fair exchange for labor, goods, or services.

Nechansky introduces computers as the controller structures in the relationship of the individuals, which differ from exclusively human-to-human interaction and decision-making [44]. Computers work with fixed goal values and defined parameters for data analysis. Modeling programs match human modeling decisions with algorithms for pattern recognition, prediction, and action decisions contextually within the hierarchy of goal values.

The interaction and agreed-upon actions may result in similar goal-value models, but this is not the same as equality or equal action outcome recommendations. Individuals remain "individual." A and B each result in individual and shared goal values whenever they cooperate in a decision for action. Individuals may be manipulated through culture or control of information, leading to domination and subornation. Current economic models based on scarcity of goods and social positions also produce constraints for shared goal values within existing social systems.

3.1.2.2 Formal Axiology

Axiology is the general label for value theory and value science [34]. Formal Axiology is a logical frame of reference or a system of value and valuation forms [34]. It is a branch of Axiology Value Science theory outlining what individual humans value (values) and how humans value (valuations). Dr. Robert S. Hartman developed the Hartman Value Profile (HVP), an empirically developed scientific approach to traditional Formal Axiology that measures the richest value to the critical destruction of an individual's choices and actions [24]. Theoretical Formal Axiology applies a parallel form through value calculous. It is the quantification of the desirability (positive or negative) of something and how it conforms to its intention.

Value is the result of the brain's conversion of direct experience (a primary, factual judgment) into meaningful truth (a secondary judgment of value) [34]. The axiom of value is grounded on the concept of fulfillment. Formal Axiology mimics human thought with three basic value types: intrinsic, extrinsic, and systemic. Value is ranked, based upon relative worth, into a hierarchy of value. Formal Axiology is applied to Business leadership, ethics, and decision-making. Formal Axiology offers neutral metrics of religion and politics; thus, diverse social and cultural approaches and solutions can interact side-by-side.

In 1989, Mefford added consideration for emotional emphasis [34]. People structure meaning through emphasis. A person makes conscious and conditioned subconscious judgments every hour and every day. Value judgment enables humans to cope, survive, and thrive in our world and society. Values unite the human capacities of rational intelligence with emotions or attitudes, which apply an emphasis [value and valuation (by emphasis) = (Knowing + Feeling)] [34]. Therefore, we value something based on our knowledge and how we feel about it. Values reveal the foundation of human character in the context of life perspective.

The transfinite cardinals, a symbolic and synthetic characteristic that effectively reflects the vastness of human emotional intensities, pose significant challenges in their interpretation into algorithms that computers can apply at scale. Forrestand has made notable advancements in formalizing a value calculus to express set theory and values mathematically [13]. The recent advancements in quantum computing hold the potential to overcome these barriers and facilitate the integration of individual values into societal automation. With the ability to handle calculus-based mathematics of Formal Axiology, quantum computers' enhanced processing capabilities could enable the incorporation of complex and nuanced personal value into automated systems. This development promises to foster a more inclusive and personalized approach to societal automation, potentially reshaping various aspects of technology and decision-making.

It is crucial to develop AI systems that can accurately recognize and respect the value of each individual, contributing to ethical and human-centered technological advancement. Formal Axiology offers a hierarchical framework for categorizing significant human value factors. If AI is involved in regulating these feedback loops, it can help steer technological progress through intentional human computation. By establishing feedback loops within society and technology and leveraging technology that integrates emerging human computation, we can advance and evolve both society and technology side-by-side.

The structural design of the information system network for sustainability data should be intentionally crafted to accommodate self-organizational contexts, utilizing implicit rules, reference points, or minimal specifications to delineate an attractive empirical form. Applying the autopoiesis theory Human Computation modeling with Complex Controller Structure, modeled with Formal Axiology provides the frame for societal automation to include sustainability to the individual level. This approach enables intricate details to emerge within the framework of self-organization, thereby facilitating adaptability while preserving equilibrium. Morgan prompts readers to critically evaluate frameworks, hierarchical systems, regulations, cultural norms, defensive mechanisms, power dynamics, and psychological pitfalls that may perpetuate the adherence to societal patterns that have become obsolete.

3.1.3 Cybernetics

In the logic of change, cybernetics focuses on mutual causality, which creates non-linear systems where loops of circular patterns of interaction are ever-evolving [40]. The use of cybernetic models of organization for institutional governance and management is not new. In 1989, Robert Birnbaum suggested how bureaucratic, collegial, political, and anarchical subsystems could operate within cybernetic paradigms. He described sensing mechanisms with positive and negative feedback loops that collectively monitor changes from acceptable levels of functioning and activate forces within processes that function as institutional thermostats [4].

Systems operate within constraints. Considering the logic of change process cybernetics, organizations can gain insights as to which slight changes can create significant impacts by considering the following questions [40]:

- Which feedback loops are significant to defining a system?
- Are there primary subsystems of loops that connect and influence one another?
- What are the primary connections?
- What are the fundamental patterns of interaction?
- Are there general forces producing these problems?
- Where is the best place to intervene?
- Are there manageable patterns to remove or add feedback loops for stabilization?

The scientific investigation takes a pragmatic, real-world philosophical view for solution-centered theory building. Systems thinking emerged as a pragmatic and well-researched domain to offer the analytic skills to understand the complexity dilemma within social systems, consider how technology may support analysis and simulations to predict behaviors and apply robust theories to devise a concept to produce desired effects. Systems thinking methods such as Soft System Methodology (SSM) are considered for workflow automation for network engagement, model building, solution design, and testing, addressing complex social challenges. System dynamics also emerged as a pragmatic and well-researched domain that offers the analytical capabilities to get to the roots of complex societal activities, introduce predictive capabilities for policy design, and create feedback loops to monitor and continuously adapt to improve social outcomes. System dynamics approaches can be applied to technology for predictive analytics of the non-linear behavior of complex systems over time using stocks, flows, internal feedback loops, table functions, and time delays [16].

Specifically, cybernetic feedback loops applying system dynamics can contribute to a quantitative approach in the following ways [60]:

- The inquiry and representation of systemic structures regarding the variables, stocks, inflows, outflows, and auxiliaries. This inquiry includes their cause-and-effect relationships and characteristics in a specific domain;
- The identification of the feedback loops and modeling of their typology. Specification of positive or negative feedback loops generated by the hierarchy of causal relationships among the variables

- The understanding of the associated dynamics with those systems through the use of system archetypes
- The exploration of the short-, medium- and long-term effects of our decisions and policies through the dynamics generated by the simulation models
- The predictive capabilities of considering which side-effects, or unintended consequences, may stem from decisions and policies
- Gaining policy insights, including learning, increases awareness about how sustainability management works by comparing forecasts against actual outcomes to improve modeling and predictive capacity over time. This information fosters awareness of complex issues related to various dimensions of sustainability (i.e., the environmental, the economic, and the social ones).

These insights may support decision-making and further development of policies for sustained improvement and catalyzation for improved social outcomes when implementing change activities [58].

Marvin Minsky is a researcher in artificial intelligence who has been researching human-centered computing for decades and points out that one must first define a specific type of problem to determine ways to tell computers to think about the problem [39]. He considers human consciousness and how the human mind simultaneously holds recollection, representation, embodiment, emotion, expression, narration, intention, apprehension, and reasoning as mental processes while pondering a problem and potential solution [38]. If computers are to assist with operational and analytic suggestions for social problem-solving, there will be different ways to represent and consider different kinds of knowledge.

Rule-based automation can address simple, expectable, or recurrent situations. For more complex problems or situations, computers can facilitate recognizing problem types to activate a way to think by first applying a critic-selector-based machine to recognize a problem type [37].

For more complex situations, a decision tree representing knowledge and skills stacks layers of methods with increasing expressiveness [37].

Training models that compare forecasts against actual outcomes improve modeling and predictive capacity over time. This cybernetic evolution fosters adaptability for complex issues related to various dimensions of sustainability (i.e., the environmental, economic, and social ones).

Complex problems may direct machine methods and knowledge-based algorithm hierarchies with layers based upon control factors such as the planetary and social boundaries layered with dependencies.

3.1.3.1 System Dynamics

System dynamics offer an approach to understanding the roots of complex behaviors to anticipate and adjust to improve outcomes. Human social systems present unique challenges to modeling human behavior. Successful modeling, development, experimental testing, and feedback loops require active engagement across various people in the policy design process, technical engineers, and the people and businesses affected by the system.

The field of system dynamics emerged at the Massachusetts Institute of Technology (MIT) in the 1950s by Jay Forrester. System dynamics and systems thinking both seek to understand and improve systems [18].

System dynamics (SD) is an approach to understanding the non-linear behavior of complex systems over time using stocks, flows, internal feedback loops, table functions, and time delays [16]. It is designed to help us learn about the structure and dynamics of complex systems. It supports developing high-leverage policies for sustained improvement and catalyzation for successful implementation and change [58].

Businesses operate within dynamic systems. System archetypes for business lifecycle are well researched and applied throughout business strategy forecasting and business intelligence [5]. Applying dynamic system modeling to an information system network for social sustainability offers a structure for incorporating mutual causality and loops of circular patterns of interaction into analytics to provide insights into the dynamics of the social systems. Applying machine learning or artificial intelligence offers the opportunity to adapt analysis continually. Over time, this holds the potential for self-adapting cyber and physical interaction to improve social outcomes and minimize unintentional impacts on the social system [43].

Strategic solutions hold interdependencies, and each planned solution creates a repercussion or unique effect on other problems. Despite the best attempts of science and technology to bring predictability and control to societal activities, the world is changing rapidly, both by intent and accident [45]. Actions taken to address one problem may lead to unintended consequences for another issue. Market and competitor reactions further influence the outcomes in unpredictable ways. Such dynamic, complex problems are referred to as wicked problems [52].

Wicked problems are problems worth solving that are difficult or impossible to solve due to their interconnected nature within dynamic and reactive systems. Wicked problems do not end. Instead, they evolve or devolve. This state of affairs is made tolerable through the awareness that desired change can be influenced by human intention and action [45]. This human intention is instrumentalized through design, which enables us to create conditions, systems, and artifacts that facilitate the unfolding of human potential in contrast to an evolution based on chance and necessity [45]. Therefore, the problems are never solved but somewhat improved, or the associated negative consequences are mitigated.

Contributions in the field of social learning and activity-theoretical approach offer a phased theory of learning that can be written into computational models and applied to wicked problems [14]. Engeström argues that expansive learning is the type of learning needed for transformations of activity systems and fields of activity [14].

Expansive learning acknowledges that learning cannot be imposed and controlled but must be implemented in a manner that allows for comparison and contrasts against the actual processes the learners perform. As with the complex dynamic nature of wicked problems, theory and reality never fully coincide [14]. Analysis articulates and bridges the gap, opening space for negotiations and contestations and putting the formation of participants' agency in the center of expansive learning. Together with a prescriptive expansive learning cycle, learning and cognition previously believed to be unique to the human mind and influenced by culture have been theoretically overcome and can be computed as "cognition in practice" [14, 28].

Although many obstacles exist when planning for dynamic, complex problems, business intelligence is moving away from spreadsheet-dominated forecasting. Dynamic modeling provides executives with the tools for testing strategic decisions in a risk-free environment. Increasingly, expansive learning is connected to rapid change over overall concepts of production, business, and organization in all spheres of economy and society [47]. By identifying feedback loops and interdependencies, simulations can be run to uncover unintended consequences, predict market reactions, and foresee delayed benefits with embedded system dynamics and expansive learning analysis for better resource planning.

Business intelligence is moving toward intelligent forecasting to provide executives with the tools for testing and directing strategic decisions. Including comparable social value metrics and data creates new visibility for strategic business decision-making, supporting investment decisions and market competitive comparative data sets. It also provides visibility for business opportunities to solve social problems profitably and improve compliance and risk management monitoring and reporting.

Generic System Archetypes for Dynamic Theory describe expected behavior patterns in organizations considering system dynamics [5]. Archetypes can be applied as a starting point structure. Over time, actions can be simulated with predictive algorithms to alert managers to unintended consequences and aid decision-making. Change activities often result in unintended side effects beyond the original goal, either positive or negative [58]. This stimulates further actions and goals. Time delays are also relevant as some side effects present immediately, while others may only appear after time has elapsed.

In 2016, a study published in the Engineering Journal generic archetypes were applied to understand the fundamental dynamics between the SDGs [67]. Applying causal loop diagrams with systems dynamics to express the relationships between SDGs and examine system structures using three generic system archetypes: Reinforcing Growth, Limits to Growth, and Growth and Under-investment [67]. This research showed that generic archetypes and causal loop diagrams can be applied beyond business as system analysis to simulate and potentially leverage points to support intentional and minimize unintentional changes in the system for sustainability.

System dynamics approaches can be applied to technology for predictive analytics of the non-linear behavior of complex systems over time using stocks, flows, internal feedback loops, table functions, and time delays [16]. System archetypes for business life-cycle are known and applied throughout business strategy forecasting and business intelligence [5]. Applying dynamic system modeling to an information system network for social sustainability offers a structure for incorporating mutual causality and loops of circular patterns of interaction into analytics to provide insights into the dynamics of the social systems. Applying machine learning or artificial intelligence offers the opportunity to adapt analysis continually. Over time, this holds the potential for self-adapting cyber and physical interaction to improve social outcomes and minimize unintentional impacts on the social system [43].

3.1.3.2 Systems Thinking

Barry Richmond, the originator of systems thinking, defines it as the art and science of making reliable inferences about behavior by developing an increasingly deep understanding of the underlying structure [51]. Systems thinking is literally a system of thinking about systems [1].

Systems thinking experts challenge established theories and models on sustainability [60]. Systems theory is uncovering the role of synergy in the evolution of living systems. The synergism hypothesis explains the cooperative phenomena of synergy by describing the collective and interdependent activities giving rise to the progressive emergence of evolutionary complex living systems [9].

As governments, commerce, and communities become increasingly interconnected, there is a growing interdependence on the technology and information systems that connect the data. As society interconnects data related to social systems, computational environments rapidly grow complex. Challenges and problems often arise as unintended consequences of well-intended past solutions.

A 2015 landscape theory review with a synthesis approach by Arnold and Wade consolidated the definitions of systems thinking into the following comprehensive description: Systems thinking is a set of synergistic analytic skills used to improve the capability to identify and understand systems, predict their behaviors, and devise modifications to them to produce desired effects [1]. These skills work together as a system.

Systems operate as continuous feedback loops that continuously improve and adapt. The critical components of systems thinking include [1]:

1. Recognizing Interconnections
2. Identifying and Understanding Feedback
3. Understanding System Structure
4. Differentiating Types of Stocks, Flows, Variables
5. Identifying and Understanding Relationships
6. Understanding Dynamic Behavior
7. Reducing Complexity by Modeling Systems Conceptually
8. Understanding Systems at Different Scales

Systems thinking is commonly misunderstood as an assessment methodology instead of a skill and worldview [31].

Bringing it all together within the context of sustainable data and cybernetics. The term sustainability presents the three well-known dimensions of sustainability: environment, economy, and society. The well known Sustainable Development Venn diagram on the represents sustainable development as only existing at the intersection of economic, social, and environmental factors. Traditional sustainability concepts consider the three dimensions of environment, economy, and society as interrelated and equal domains and decision-making processes.

A recent interpretation considers the inter-relatedness of the dimensions and offers a hierarchy of dependence. Systems thinking stipulates that an economy is dependent upon a society, which depends on an environment. The environmental dimension is a requirement for the others, with limits and possible points of no return, or failure. Further, the societal dimension is regarded as a requirement for an economy, as there must be a marketplace of individuals and rules for exchange and order. Society also has societal limitations.

When applied to technology, system dynamics approaches may offer predictive analytics of the non-linear behavior of complex systems over time using stocks, flows, internal feedback loops, table functions, and time delays [18]. They offer insights into interaction patterns, forces that influence and produce problems, and significant feedback loops, which can act as predictive factors within a system over time. This offers insight into the best time and place to intervene for stabilization and improved outcomes. Such system analysis of SDGs can support targeted decision-making and provide insights into potential social value/social impact for societal automation. Although there are many obstacles when planning for dynamic complex problems, dynamic modeling provides the tools for testing strategic decisions in a risk-free environment. By identifying feedback loops and interdependencies, simulations can uncover unintended consequences, predict market reactions, and foresee delayed benefits for better resource planning. It supports the design and development of high-leverage policies for sustained improvement and catalyzation for successful implementation and change [58].

These process steps can be machine-augmented with learning loops and workflow automation to enhance stakeholder engagement and problem-solving within complex social systems where stakeholders hold diverse worldviews for societal automation.

Human social systems present unique challenges to modeling human behavior. Businesses operate within dynamic systems. Strategic solutions have interdependencies, and each planned solution creates repercussions or influences other problems. System dynamics offer an approach to understanding the roots of complex behaviors to anticipate and adjust to improve outcomes.

Cybernetic feedback loops applying system dynamics can contribute to a quantitative approach in the following ways [60]:

- The inquiry and representation of systemic structures regarding the variables, stocks, inflows, outflows, and auxiliaries. This inquiry includes their cause-and-effect relationships and characteristics in a specific domain;
- The identification of the feedback loops and modeling of their typology. Specification of positive or negative feedback loops generated by the hierarchy of causal relationships among the variables

- The understanding of the associated dynamics with those systems through the use of system archetypes
- The exploration of the short-, medium- and long-term effects of our decisions and policies through the dynamics generated by the simulation models
- The predictive capabilities of considering which side-effects, or unintended consequences, may stem from decisions and policies
- The gaining of policy insights, including learning, increases awareness about how sustainability management works by comparing forecasts against actual outcomes to improve modeling and predictive capacity over time. This information fosters awareness of complex issues related to various dimensions of sustainability (i.e., the environmental, the economic, and the social ones).

These insights may support decision-making and further development of policies for sustained improvement and catalyzation for improved social outcomes when implementing change activities [58].

Information artifacts from individuals act as agents for cyber-physical coordination to trigger meta-processes within the environment. Marvin Minsky considers human consciousness and how the human mind simultaneously holds recollection, representation, embodiment, emotion, expression, narration, intention, apprehension, and reasoning as mental processes while pondering a problem and potential solution [38]. If computers are to assist with operational and analytic suggestions for social problem-solving, there will be different ways to represent and consider different kinds of knowledge.

Training models that compare forecasts against actual outcomes improve modeling and predictive capacity over time. This cybernetic evolution fosters adaptability for complex issues related to various dimensions of sustainability (i.e., the environmental, economic, and social ones).

Complex problems may direct cybernetics and machine methods with knowledge-based algorithm hierarchies organized in layers based upon control factors such as the planetary and social boundaries linked with dependencies. Considering the Three Dimensions of "Sustainability" and "Sustainable Development" and the more recent interpretation of dependencies between the three dimensions. Establishing a hierarchy that considers environmental stability a priority over social stability over the economy, reasoning that there is no economy without a society and no society without an environment.

3.1.4 Change as the Tension Between Opposites

As noted in cybernetics, system dynamics result in circular loops. Each action stimulates a counteraction and often unintended consequences within the system. The Generic Archetypes of Dynamic Theory represent the most common patterns observed in organizations [5]. These diagrams describe the repeated patterns organizations must navigate resulting from the phenomenon of opposites within our economic system.

Morgan's metaphor in Flux and Transformation further explores the process of change, with change as the tension between opposites. This concept looks deeper to build awareness of differing perceptions that stimulate its opposite effect when change occurs.

To begin, consider the term opposite and that measuring a state of social value requires defining the opposite state of social destruction. As a basic example, consider the limits; we consider day and night, good and bad, happy and sad, positive and negative. The ability to understand and define the existence of one state requires the ability to define what is not within the scope of that state. Opposites exist in a state of tension within a dynamic and changing system.

Generations of social theorists and scientists have pondered this paradox. Taoist notions of Yin and Yang seek balance in opposing elements. In the Book of Changes, Taoist I Ching applies archetypal patterns found in nature and community to formulate a way of thinking about opposites. There is a rich and complex nature within the process of change.

The interplay of opposites is as alive today as it was centuries past, perhaps even more today in the rapid transformational times of technological change filled with economic and social contradictions. Social theories also consider the tensions of opposites on an individual and societal scale. Morgan focuses on the metaphor using an oversimplified example of the tensions and contradictions of capitalist employment to describe how the interplay of opposites fuels social change and transformation. The very act of change implies the destruction of what is today in some way to bring about improvements for tomorrow.

The tensions between opposites acknowledge that today's decisions will inevitably defeat their intended purposes. The example of tension and contradiction reminds us that there is a predictable negation and counter-negation of progressive development within a changing social system despite good intentions.

Globalization has brought an increase in power to multinational corporations [21]. The classic economist's amoral perspective on the purpose of business is to maximize profits within legal limits [19]. In this business context, leaders, specifically of publicly held companies, may not allow moral standards to divert from the goal of maximizing profits for the shareholders. Alan Goldman points to the moral justification of profit maximization by focusing on the "within legal limits" portion of this definition [15]. He argues on utilitarian grounds that public needs and wants are best satisfied in a profit maximization model due to the legal context of "rights." A free-market economy based on profits best serves to honor the "rights" of consumers due to their ability to choose between products and companies.

When considering social sustainability from a business perspective, the issue of economics is front and center in a manner that institutions often disregard or underestimate. It is outside the perspective of government and institutions to stress the necessity for business to link their social value contribution to business economics. It is within reason that those political leaders prioritize a business's social and environmental outcomes. Policymakers care less about whether a business is making money or whether it costs the business to activate or achieve those outcomes. In comparison, the business considers economic factors a higher survival priority.

Businesses that attempt to balance economic and social considerations find the need for systemic data and processes for measuring social value an inhibitor. Only some things that offer social value generate revenues, much fewer profits. Through economic considerations, businesses can allocate resources with more information to make strategic decisions about where to invest or which activities to replicate across markets [20].

Much discussion and controversy exist regarding business measurements of social responsibility and good governance, particularly ESG standards. ESG standards are rising in importance and frequency of use concerning business investment reporting [59]. Organizations are reporting year-on-year with increasingly comprehensive criteria. However, studies show inconsistencies across organizations, sectors, and regions. ESG ratings consistency and comparability of data are poor. The reasons identified are lack of disclosure transparency, company size bias, geographic bias, industry bias, inconsistencies between rating agencies, and failure to identify risks [12]. Other forms of organization, such as associations or non-profit organizations, also balance the economic and social impact goals. Even though some forms of the organization hold a less economically driven goal, they are also economically bound for survival.

Perception bias falls into the category of tension between opposites. It is when a firm states and believes that it has helped people, but. in the end, those who were helped may report that there was no realized value. For example, a firm might report creating 100,000 jobs in society as a positive impact. However, with further investigation, the wages for those jobs are below a living wage, so those people will not live to their full potential. Employees need help to afford to meet their needs, yet a large corporation employs them. The business is often not aware of its bias.

Firms may cherry-pick the results that help market their contribution to their customers, markets, or investors [11]. Perception bias may result in positive marketing or the damaging consumer accusation of greenwashing. A business has a survival motivation linked to economic factors. Although they operate with respect to regulations and laws, they also have the motivation and incentive to exploit variations in requirements across different environments. The endurance of society does not necessarily exclude exploitation.

Governments seek to facilitate transparency and accountability that benefits citizens—for example, reducing the cost of corruption to create a level playing field by opening the tender process and procurement contracts. However, there are limitations to control mechanisms at a global scale as large businesses operate internationally with inconsistent accountability structures, which allow for the displacement of risk.

Some business researchers acknowledge the tensions between opposites. According to Porter, pure capitalism and a business's profit orientation can have exploitative effects on social structures. Measuring business results has historically taken a profit-focused approach with financial models that assume infinite growth on a finite planet. Social responsibility measures have begun to track how a business decision affects other individuals, groups, or environments.

Governments, NGOs, and philanthropic organizations seek to fill social gaps under-served by market forces. However, there is an issue of scale [48]. NGOs, governments, and philanthropic organizations do not have enough financial resources to address today's challenges, e.g., global warming, social inequality, and disaster management.

Porter suggests reinventing capitalism in a manner that directs businesses to tackle social issues [3]. He recommends that the purpose of business be redefined beyond profit to create shared value. To support this goal, he collaborates to develop strategic models to guide corporate policies and practices that enhance a company's competitive advantage and profitability while simultaneously advancing social and economic conditions in the communities it sells and operates [3]. Porter argues that shared value is not corporate social responsibility, philanthropy, or sustainability. Instead, it is a new business strategy to achieve shared success.

Shared value aligns with human development theories that forecast human cooperation to develop with similar systems of organization [57]. Stagish developed the Human Development Stairway to Social and Cognitive Transformation, which represents the steps of social evolution through increasing synergic cooperation.

The Human Development Stairway to Social and Cognitive Transformation describes five steps from low synergy collaboration n to high synergy social collaboration. The lowest synergy form of collaboration is focused on basic needs: Self-centered, Aggressive, and Competitive. The stairway progresses one step up to include functional and social groups: Social/Cooperative. The third step up is understanding one's own learning needs in a social and cognitive context: Self-directed. The fourth step is collaboration, when members can affirm their own and others' identities and needs: Self-actualization. Finally, the highest synergy step of social collaboration is high-order thinking and sharing: Synergistic [57].

Shared value proposes that business moves up the stairway to a higher synergy as shared value collaborates in society to drive the next wave of innovation and productivity growth in the global economy [48]. As communities become more efficient and organized into functional cooperative social groups, they gradually interact in increasingly synergistic self-directed systems. There are evolving approaches to Shared Value that consider societal impact while resulting in productivity benefits. Shared value can also capture a complete set of societal indicators, environmental sustainability, and compliance data for external reporting. As a data-focused model, qualitative and quantitative data are refined for business intelligence. This information may offer new opportunities to identify partnerships with complementary expertise and serve social needs within an ecosystem [20].

Note that potential new futures always create new tensions from the opposition to the status quo [40]. Innovation can lead to an established practice's evolution and destruction (of the old practice). Morgan suggests considering primary and secondary contradictions. Causal loop diagrams can map the initial expectation with machine learning, revising models with actual trends and results.

The dialectic analysis applied within system dynamic modeling provides a powerful lens to understand the logic of change in our past, present, and future. The paradoxical tensions between the status quo and alternative future states hold the potential to create monumental transformations.

Organizations often do not understand the complexity and recursive loops they depend upon for survival. An international or global organization can become powerful enough to achieve independence from a specific national societal system and adapt for its survival by transcending its dependence upon one country or society for its renewal. A concept for social value modeling within societal automation should consider ways to address this known issue.

Society 5.0 is the first real-world project example of a fully integrated digital social system incorporating a service economy, responses to climate change and SDGs, and slowing global economic growth. The Japanese government's Society 5.0 project seeks to create a human-centered balance for financial advancement in alignment with resolving social problems through a system that highly integrates cyberspace and physical space [6].

Individual goals, business objectives, and larger political objectives interact and contradict each other. Each situation and perspective leads to a different interpretation of the same information, resulting in the challenge of various definitions and low-quality data for social value assessments. There are multiple views of reality or other world views based upon the individual or group's position when considering the impacts of change activities.

To address these tensions of opposites, Society 5.0 incorporates a seven-step process to include human activity systems' different worldviews before making improvements [61]. They consider the unique perspectives of different roles and apply systems thinking methodologies that seek to bring in and consider these different views before taking any action towards a solution [61]. The seven-step process applies a system's thinking Soft Systems Methodology, which seeks to bring in and consider these different views through engagement, model building, and testing before taking action toward a solution.

3.1.4.1 Systems Thinking Soft Systems Methodology

Systems thinking, in particular, offers a qualitative approach to system dynamics, quantitative feedback loops, and algorithm hierarchies that will be required to automate technology.

Technology creates the entry point for citizen engagement, allowing us to get closer to understanding what individuals and groups want and need. Clustering input from different groups by actor roles provides value trends and visibility of indirect impacts. SSM CATWOE is applied to define primary role groups into Customers, Actors, and Owners. Policymakers can combine the synthesis of individual and community values to gauge social pressures and demands.

The SSM iterative approach builds models for managers, researchers, and policymakers to gather information through social consultation for large-scale activities before a project is launched and is successfully being applied within the Society 5.0 initiative [61].

The following steps outline each phase of SSM implementation: 1. Identify the challenging situation, 2. Communicate about the problem situation, 3. Establish root definitions for the system (apply CATWOE) 4. Create a model to represent the concept, 5. Compare conceptualizations to real-world models 6. Evaluate possible improvements 7. Create an action plan [7].

In the final stage of SSM, improvements are implemented, and the outcomes of previous steps are evaluated. Actions are taken to address the problem situation, aiming to meet system requirements and fulfill the goals of key stakeholders and owners. If the solutions fail to meet the required levels of efficiency and effectiveness, the SSM process can be restarted with alternative approaches to solving the problem situation.

Subject matter experts, such as academia and IGOs, are structured to guide model building and prototyping. Models are tested with the communities anticipated to be impacted through ecosystem platforms such as the social, policy, and governance system (SES) platform modeled by McGinnis et al. [33] which supports policymakers' academic analytical, diagnostic, and prescriptive capabilities in a separated environment.

Cross-disciplinary teams and groups consider the interconnections which combine or create tensions, to build cause-effect dynamic feedback loops. Dynamic modeling embeds predictive capabilities for policy design. Feedback loops are designed to monitor and report for continuous adaptation potential to improve social outcomes.

Engagement across scales enables transparent and inclusive feedback loops within a minimum specification through automated workflows. Such engagement across scales allows for reframing stabilizing feedback loops to provide increased knowledge of patterns across society to support the harmonious equilibrium of relationships within and across the social system.

Critics have a good point and note that SSM assumes that involved parties intend to cooperate and reach a consensus. It acknowledges that those with power have a disproportionate opportunity to manipulate outcomes in their favor, as in our current systems [49]. This metaphorically aligns with the logic of change "Change as the Tension Between Opposites." Which calls to question if an alternative methodology will be required for modeling.

3.2 Applying Thought Trials to Modeling

Chapter 3 serves as a critical bridge between the foundational concepts established in Chap. 2 and the practical theory building to take place in Chap. 4. By delving into metaphor-driven contextualization and the use of metaphors from Gareth Morgan's "Images of Organization" to conceptualize complex sustainability dynamics, it lays the groundwork for mentally reframing how the requirements from Chap. 2 can be applied in the model building of Chap. 4. Furthermore, this chapter integrates the relationships between systems and their environments, chaos and complexity theory, cybernetics, and the tensions between opposites to establish a solid foundation for the model presented in Chap. 4. The constant comparison and visual representation of stakeholder patterns, purpose, and challenges have paved the way for a pragmatic technical solution that aligns with the overarching goal of harmonizing social sustainability data across intelligent information networks.

Systems Thinking together with Systems Dynamics synergize activities for cyber-physical coordination. System dynamics offer analytical capabilities to get to the roots of complex societal activities, introduce predictive capabilities for policy design, and provide feedback loops to monitor and continuously adapt to improve social outcomes.

Figure 3.1 shows the operations and analysis by scale diagram for system dynamic structuring. It addresses the general requirements from Building Block I and gives considerations for the paradox of eternal change and dynamic complexity:

Mutual causality creates non-linear systems where loops of circular patterns of interaction evolve [40].

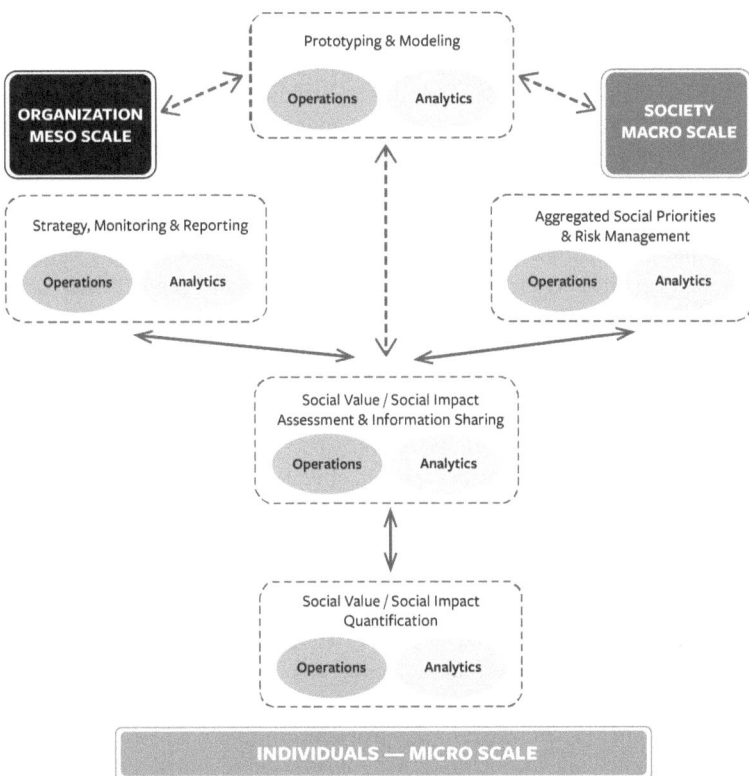

Fig. 3.1 Operations and analytic by scale diagram for system dynamic structuring

The Information System Network for Social Sustainability operates within and analyses data from feedback loops across a societal network in a manner that continuously adapts over time and can process data from different perspectives of scale. Cybernetic methodologies for dynamic systems with mutual causality use feedback loops and interdependencies across operations and analytics. Data from various perspectives capture value and negative impacts to understand, influence, and balance social tensions.

The Conceptual Model aims to synergize and promote harmony across known patterns of contradictions that shape our society and life. Knowledge of the circular loops of interaction patterns supports identifying practices that improve outcomes. The feedback looping models apply stocks, flows, internal feedback loops, table functions, and time delays [16]. System archetypes further provide the foundational patterns within organization systems. The dialectic analysis applied within system dynamic modeling provides a lens to visualize and understand the effects of change in our past, present, and future.

Semantic platforms with data loaders, sorting, labeling, linking, machine learning, dashboards for visualizations, and broader social information network sharing offer hubs for information aggregation and sharing [54]. To develop a semantic information system network, precise definitions, classifications, characteristics, typing, and tiering are required for machine learning algorithms to be trained.

Primary data labels are required for machine-sortable decision-making as listed in Chap. 2 by Maas and Likert. This meets Building Block I Requirement nr.4 , noted in Chap. 1, needed to allow technology and platforms to facilitate semiotic analysis and processes with comparative-quality data across all scales.

The application of systems thinking's analytical skills for synergizing activities is prominently observed in the modeling process for societal automation leveraging collective intelligence. Systems thinking is a set of synergistic analytic skills used to improve the capability to identify and understand systems, predict their behaviors, and devise modifications to them to produce desired effects [1]. These skills work together as a system.

Stigmergic Systems emerged as a scale-free architecture for self-organization processes to leverage collective intelligence in societal automation. These systems offer the opportunity to frame groups of software services based on proposed fields of activity to enable appropriate storage and operating solution stack configurations for service deployment.

Setting the specific boundary conditions for social stability is beyond the scope of this research. Nevertheless, the model design should consider the implementation of such criteria. From a practical view, Dr. Kate Raworth's doughnut of social and planetary boundaries provides an example of a practical starting point for social dimensions defining potential limits to social stability [53]. Raworth's theory also offers flexibility for activities within these social and planetary boundaries, allowing for an environmentally safe and socially just space for humanity to thrive.

Raworth's framework can be effectively utilized in technical contexts by examining the participation model within stigmergic information systems. These systems enable the intricate behaviors of communities through software architectures

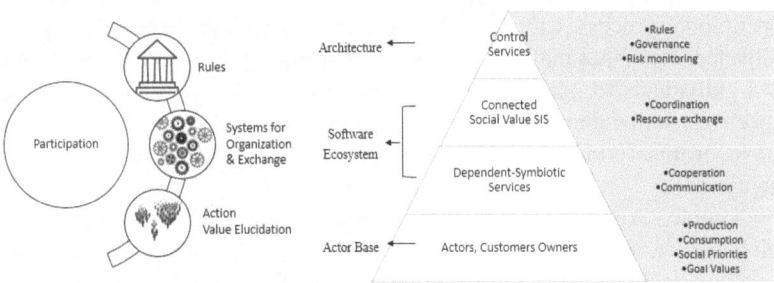

Fig. 3.2 Pyramid of participation in a stigmergic information system network for social sustainability

specifically engineered for self-organization and automated data capture, thereby streamlining interaction processes. This entails deploying control services for establishing rules and governance, the creation of systems for coordinating organization and facilitating resource exchange, and carefully considering individuals in terms of action and value elucidation.

Drawing from chaos and complexity theory, when the dominant attractor pattern allows for an undesirable state, the key challenge lies in fostering sufficient flexibility to facilitate the emergence of new behavioral patterns. As machine learning models continue to advance, a stigmergic social sustainability information system network has the potential to provide recommendations and guide autonomous corrective actions and activities aimed at promoting social sustainability.

Figure 3.2 aligns the stigmergic information system pyramid of participation from the SIS and applies systemic social sustainability aspects. The Stigmergic Information System Network for Social Sustainability is designed with a minimum set of stability-focused criteria that facilitate the flexible, flowing process of change that can endure rule and control service adaptations as society evolves.

This base structure is flexible enough to allow for various scales, role perspectives, boundary conditions, and processes. Ant Colony Optimization (ACO) from Swarm Intelligence can be applied to simple computational modules to direct multiple individuals to contribute to a collective goal of model building.

The Stigmergic Information System shares and applies social value and social impact data across the ecosystem, and semantic platforms cooperate to link and share social value, shared value, and social sustainability information across the Stigmergic Information System network.

By systematically structuring data and leveraging intelligent data tools to separate personal data to retain privacy while working with dual operational and analytic tools and environments, cleaned relevant data can be organized to be externalizable and shared through service interfaces. Distributed programmed modules or components can be reused, updated, and share data for monitoring and adapting.

Technology provides a vehicle for role grouping and workflow automation of operational processes to drive essential information exchange through swarm intelligence. Processes may be operational or analytical at all scales. Role labeling allows individuals to perform tasks and provide social information from various perspectives. Individuals provide quantification of social value, both positive and negative, of transformation activity. This can be conducted through the automation of engagement.

For simple, expectable, or recurrent situations, rule-based automation can be developed. For more complex problems or situations, computers can facilitate recognizing problem types with a critic-selector-based machine to activate more complex layers of ways to think [37]. The more complex the social situation, the more layers of ways to represent knowledge and skill may be arranged in a stack with increasing degrees of expressiveness by applying semantic networks, neural networks, K-lines K-trees, etc. [37]. Applying various decision methods, such as machine learning or artificial intelligence, offers the potential for self-adapting cyber and physical interaction to improve social outcomes and minimize unintentional impacts in the social system [43].

3.2.1 Perspective Matters When Collecting and Sharing Social Value and Social Impact Data

Assessment has to do with an evaluation. Perception considers our understanding of the relationship between factors as a whole. How the assessor views the problem requires a perspective. This subjective point-of-view can hold notable significance when measuring social value and social impact on social sustainability. Stakeholders each assess for a reason and from their perspective.

In Fig. 3.3, the Social Information Flow Overview reduces the model into perceptual wholes, stripping out secondary details to reduce complexity. It distinguishes different scales within the system and indicates interactions and connections. It diagrams how the structures influence systematic behaviors. Boundary conditions related to each scale are inside parenthesis. Primary relationships of information flow for systemic alignment are noted with arrows.

Technology mediates the measurement and provides analysis of social aspects of activities. Technology initiates micro-tasks of self-organization processes within complex systems to promote equilibrium and stable states. Stability is not exclusive to change and transformation but instead, the opportunity to implement a form of regularity through change [25].

The foundational structure is a minimum set of criteria that facilitate society operating with limitations for social stability while allowing the flowing processes for unrestricted dynamic activities that fall within social sustainability boundary limitations.

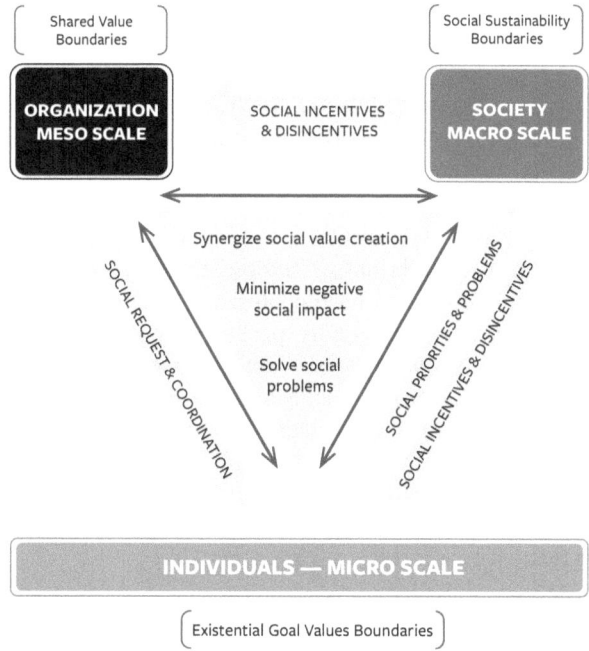

Fig. 3.3 Social information flow overview

Boundary conditions are programable parameters that technology can apply as selection criteria for analysis to ensure social sustainability.

Semiosis is a part of an ongoing process of communication. It is a chain of triads with a focal factor of a dynamic process [2]. A signal initiates an interpretation or condition at a focal level. This interpretation or condition is processed within limitations or boundary conditions.

Semiotic behaviors stabilize communication with a system through self-correcting dynamics of looping interconnections [46]. New behaviors emerge from previously existing cultural evolutionary processes. Stability and the tendency to develop new systemic interaction habits become stabilizing forces and develop autonomously [2].

The model differentiates scales, where each living system is characterized by autonomy, circularity, and self-reference with an ability to self-create and self-renew. Consideration of the unique boundary conditions of scale is necessary.

3.2.1.1 Macro Scale

The boundary condition for the macro scale is social sustainability. Social sustainability is the capacity of human society to endure. Social sustainability interconnections refer to the formal and informal processes, systems, and relationships for stable communities today and for future generations. Dr. Kate Raworth's "Doughnut" of social and planetary boundaries offers the floor and ceiling boundary conditions that combine the nine planetary environmental boundaries as an example of boundary conditions [50]. Its twelve dimensions are derived from the SDGs 2015. Raworth proposes that activities within these social and planetary boundaries allow for an environmentally safe and socially just space for humanity to thrive.

Policies added to control service modules for algorithms of incentive and disincentives can act as macro attractors to pull the societal system into states of equilibrium (or near equilibrium). Activities that do not jeopardize breeches in the boundary limits are not to have control rules at the macro scale. Despite the unpredictability, a systemic gravity pushes or pulls for spontaneous self-reorganization, whereas a coherent order always emerges from the randomness [27]. Technology supports decision-making to avoid destabilization and aids in self-correction with predictive guidance, suggestive decision-making, and retrospective analysis.

3.2.1.2 Meso Scale

Organizations include cooperative groups. Not all are in the form of a business, nor are they all profit driven. The boundary conditions are unique as an organization is monitored to ensure that it does not breach the boundary limitations of its Macro environment nor the boundary conditions of existential goal-values of the Micro participants of their activity. Micro participants include all owners, actors, and customers (direct and indirect) of group activities. For the self-sustainability of a group/organization, the boundary condition to create or produce a net shared value is set for computer-aided decision-making algorithms. Shared value recognizes that an organization seeks to generate both net social benefits and benefits for the group. This view also acknowledges and incorporates the quantification of social harms, or destruction, within the net shared value quantification. Therefore, these costs, in the form of social risks, can be carefully managed. Shared value considers the economic factors of the organization, existential goal-values of participants, and the social sustainability of its environment to synergize the potential for all to endure.

Algorithms are programmed to offer analysis and suggestions that meet the organization's shared value boundary conditions and verify operation within the macro scale boundary limitation. Organizations are hierarchically below that macro-scale boundary conditions for algorithmic decision-making. Organizations are also hierarchically bound by rules associated with not intentionally breaching the individual existential goal-values, micro boundary conditions. Organizations are guided to transform or dissolve if they do not bring about a net social value.

3.2.1.3 Micro Scale

The quantification of social value/social impact from the individual perspective at the micro-scale is a new opportunity that scale-free distributed technology introduces.

Applying the autopoiesis theory Human Computation modeling with Complex Controller Structure, modeled with Formal Axiology provides the frame for societal automation to include sustainability to the individual level. Boundary conditions for the individual are set as existential goal values. A human does not live forever, and death is currently a normal part of the human lifecycle. Therefore, calculations must be within the scope of a human lifetime and perspective. Formal Axiology is a feasible example of a well-developed domain of science to quantify the existential goal-value boundary limit for the individual [24, 34].

Short-term and long-term goal definitions are dynamic and ever-changing algorithmic variables set by the individual. The short-term and long-term individual goal values are unique, with variations based upon shared goal values when an individual plays a role within a group. Different algorithms are applied based upon the role of the actor, customer, or owner concerning the initiating event.

The Complex Controller Structure, proposed by Helmut Nechansky, provides a valuable framework for modeling human decision-making processes and behavior. Within an information system network, the Complex Controller Structure is utilized at the individual level to design and implement computer-based decision-making systems that consider the epistemological aspects of individual decisions and behaviors. By incorporating aspects such as goal-value setting at different levels and utilizing machine learning models for continual adaptation, the system can be designed to analyze historical sensor data, identify patterns, and align with long-term and short-term goal values.

Furthermore, the application of Formal Axiology, also introduced in Chap. 3, within the information system network can provide valuable insights into value theory and valuation forms related to societal automation. Dr. Robert S. Hartman's development of the Hartman Value Profile offers an empirically developed scientific approach to traditional Formal Axiology, providing a framework for measuring the value and impact of individual choices and actions. The quantification of desirability and conformity to intention in value science theory can be harnessed to assess and analyze societal data within the information system network. Moreover, by considering emotional emphasis in value judgment, the system can integrate human capacities of rational intelligence with emotions and attitudes, creating a holistic approach to value assessment within the societal automation framework that ensures a comprehensive understanding of individual human value and values.

Overall, incorporating the Complex Controller Structure and Formal Axiology within an information system network for social sustainability data facilitates the development of a sophisticated framework capable of incorporating individual input at scale to analyze societal data, automate processes, and promote social sustainability by offering guidance to align individual and shared goal values. Moreover, it offers a nuanced understanding of human value and diverse social and cultural approaches, thereby contributing to leveraging collective intelligence for societal automation.

In closing, Chap. 3 provides a valuable sensory perception process enabling readers to begin to image societal automation leveraging collective intelligence as a possibility. This chapter serves as a critical bridge between the vast literature on social impact and social value data with concepts around the area of societal flux and change within computational scenarios for application. A model for unifying social sustainability data must first consider how the data will be applied and where it can be captured and shared—purposefully. By delving into metaphor-driven contextualization and the use of metaphors from Gareth Morgan's "Images of Organization" to conceptualize complex sustainability dynamics, we have laid the groundwork for the beginning of an intuition of societal automation and the role of collective intelligence in driving a sustainable future. Furthermore, this chapter integrates the relationships between systems and their environments, chaos and complexity theory, cybernetics, and the tensions between opposites to establish a solid foundation for the model presented in Chap. 4. The constant comparison and visual representation of stakeholder patterns, purpose, and challenges have paved the way for a pragmatic technical solution that aligns with the overarching goal of harmonizing social sustainability data across intelligent information networks. As an independent publication of scholarly work, this chapter contributes to the literature on sustainability, collective intelligence, and societal automation. It provides a nuanced exploration of the need for social information across information system networks.

In conclusion, this chapter has explored the use of metaphorical reasoning as a tool for comparison and abstraction in the context of reordering sustainability information for collective intelligence. By leveraging the metaphor of "flux and transformation" from Gareth Morgan's "Images of Organization," we have gained valuable insights into the complex dynamics of social value and sustainability. The metaphor has allowed us to conceptualize the interconnectedness, adaptability, and continual change inherent in social development. Furthermore, by considering the interactions within the context of "living" and the environment, we have highlighted the significance of environmental changes as primary influences for adaptation and change. The exploration of metaphorical reasoning has provided a valuable framework for understanding and representing the complex dynamics of sustainability information. By engaging in disciplined imagination and thought trials, we have used metaphor as a valuable tool to focus attention and build abstract yet generalizable sensory representations. This approach has allowed us to identify and map the relationships provisionally, paving the path for an early theoretical story to emerge. Going forward, the insights gained from this exploration will serve as a foundation for further research and practical applications in the field of sustainability and collective intelligence. By embracing metaphorical reasoning as a valuable tool for comparison and abstraction, we can continue to develop pragmatic technical solutions to meet the evolving requirements of social sustainability data across intelligent information networks. This chapter lays the groundwork for future endeavors to harness metaphorical reasoning for collective intelligence and sustainable development.

References

1. Arnold RD, Wade JP (2015) A definition of systems thinking: a systems approach. Proc Comput Sci 44:669–678
2. Atã P, Queiroz J (2019) Emergent sign-action. classical ballet as a self-organized and temporally distributed semiotic process. Eur J Pragmatism Am Philos 11(XI-2)
3. BBC Radio 4 (2011) A new capitalism. Retrieved June 6, 2011, from BBC, In Business. http://www.bbc.co.uk/iplayer/console/b00xj0r4/In. Business A New Capitalism
4. Birnbaum R (1989) The cybernetic institution: toward an integration of governance theories. Higher Edu 18(2):239–253
5. Braun W (2002) The system archetypes: the systems modelling workbook. https://www.albany.edu/faculty/gpr/PAD724/724WebArticles/sysarchetypes.pdf
6. Cabinet Office (2021) Society 5.0. Retrieved from Government of Japan. https://www8.cao.go.jp/cstp/english/society50/index.html
7. Checkland P, Poulter J (2020) Soft systems methodology. In: Systems approaches to making change: a practical guide. Springer, Berlin, pp 201–253
8. Cornelissen JP (2005) Beyond compare: metaphor in organization theory. Acad Manag Rev 30(4):751–764
9. Corning PA (2014). Systems theory and the role of synergy in the evolution of living systems. Syst Res Behav Sci 31(2):181–196
10. Davidsen J, Ebel H, Bornholdt S (2002) Emergence of a small world from local interactions: modeling acquaintance networks. Phys Rev Lett 88(12):128701
11. Dillenburg S, Greene T, Erekson OH (2003) Approaching socially responsible investment with a comprehensive ratings scheme: total social impact. J Bus Ethics 43:167–177
12. Doyle TM (2018) Ratings that don't rate: the subjective world of esg ratings agencies. American Council for Capital Formation, pp 65–71
13. Edwards RB (1995). Ten unanswered questions. In: Formal axiology and its critics. Rodopi, Leiden, pp. 145–152
14. Engeström Y (2015) Learning by expanding. Cambridge University Press, Cambridge
15. Falikowski AF (1990) Moral philosophy: theories, skills, and applications. Prentice Hall, Englewood Cliffs
16. Fleming E (2020). What is a system dynamic model? Retrieved from SidmartinBio Wide base of knowledge. https://www.sidmartinbio.org/what-is-a-system-dynamic-model/
17. Floreano D, Mattiussi C (2008) Bio-inspired artificial intelligence: theories, methods, and technologies. MIT Press, Cambridge
18. Forrester JW (1994) System dynamics, systems thinking, and soft or. Syst Dyn Rev 10(2–3):245–256
19. Freidman M (1962) Capitalism and freedom. University of Chicago Press, Chicago
20. FSG Impact (2013) Measuring shared value. Webinar, united states of america. Retrieved from https://youtu.be/RYuZcAVl0g
21. Fuchs D (2007) Business power and global governance. Lynne Rienner Publishers, Boulder
22. Garmestani AS, Allen CR, Gunderson L (2009) Panarchy: discontinuities reveal similarities in the dynamic system structure of ecological and social systems. Ecol Soc 14:12pp
23. Grassé PP (1959) Un nouveau type de symbiose: la meule alimentaire des termites champignonnistes. Nature 3293:385–389
24. Hartmann RS (2018) Robert s. hartman institute. Retrieved from Hartman Value Profile. https://www.hartmaninstitute.org/about/hartman-value-profile/

25. Kelso JS (1995) Dynamic patterns: the self-organization of brain and behavior. MIT Press, Cambridge
26. Kennedy J (2006) Swarm intelligence. In: Handbook of nature-inspired and innovative computing: integrating classical models with emerging technologies . Springer, Boston, pp 187–219
27. Langton CG (1990) Computation at the edge of chaos: phase transitions and emergent computation. Phys D Nonlinear Phenom 12(1–3):12–37
28. Lave J (1988) Cognition in practice: mind, mathematics and culture in everyday life. Cambridge University Press, Cambridge
29. Lorenz EN (1963) Deterministic nonperiodic flow. J Atmos Sci 20(2):130–141
30. Macy MW, Willer R (2002) From factors to actors: computational sociology and agent-based modeling. Annu Rev Soc 28:143–166
31. Mahmoudi H, Dorani K, Dehdarian A, Khandan M, Mashayekhi AN (2019) Does systems thinking assessment demand a revised definition of systems thinking? In: The 37th international conference of the system dynamics society
32. Maturana HR, Varela FJ (1991) Autopoiesis and cognition: the realization of the living. Springer, Berlin
33. McGinnis MD, Ostrom E (2014) Social-ecological system framework: initial changes and continuing challenges. Ecol Soc 19(2):12pp
34. Mefford DL (1989) Phenomenology of man as a valuing subject. the university of tennessee. Retrieved from https://www.proquest.com/openview/6753b29b844caf631a9dc2bac07773e6/1?pq-origsite$=$gscholar&cbl$=$18750&diss$=$y
35. Michelucci P (2013) Handbook of human computation. Springer, Fairfax
36. Minsky M (1988) The society of mind. Simon, New York; Schuster
37. Minsky M (2007). The emotion machine: commonsense thinking, artificial intelligence, and the future of the human mind. Simon, New York; Schuster
38. Minsky M (2011) Interior grounding, reflection, and self-consciousness. In: Information and computation: essays on scientific and philosophical understanding of foundations of information and computation. World Scientific, Singapore, pp 287–305
39. Minsky M (2019) A framework for representing knowledge. de Gruyter, Berlin
40. Morgan G (2006). Images of organization. Sage, Thousand Oaks
41. Mulgan G (2010) Measuring social value. Stanford Soc Innov Rev 8(3):38–43
42. Musil J, Musil A, Winkler D, Biffl S (2012) A first account on stigmergic information systems and their impact on platform development, in Proceedings of the European conference on software architecture WICSA/ECSA 2012 companion volume, pp 69–73
43. Musil A, Musil J, Weyns D, Bures T, Muccini H, Sharaf M (2017) Patterns for self-adaptation in cyber-physical systems. In: Multi-disciplinary engineering for cyber-physical production systems. Springer, Berlin, pp 331–368
44. Nechansky H (2013) Epistemological issues in human computation. In: Michelucci P (ed) Handbook of human computation. Springer, New York, pp 71–81
45. Nelson HG, Stolterman E (2014) The design way: intentional change in an unpredictable world. MIT Press, Cambridge
46. Peirce CS (1958) Collected papers: science and philosophy and reviews, correspondence, and bibliography, vol 5. Belknap Press of Harvard University Press, Cambridge
47. Pihlaja J (2005) Learning in and for production: an activity-theoretical study of the historical development of distributed systems of generalizing. Helsingin Yliopisto, Helsinki
48. Porter ME (2018) Harvard business school. Retrieved from Creating Shared Value Explained https://www.isc.hbs.edu/creating-shared-value/csv-explained/Pages/default.aspx
49. Ramage M, Shipp K (2009) Systems thinkers. Springer, Berlin
50. Raworth K (2012) A safe and just space for humanity: can we live within the doughnut? Oxfam, New Delhi
51. Richmond B (1994) System dynamics/systems thinking: let's just get on with it. Syst Dyn Rev 10(2–3):135–157

52. Rittel HW, Webber MM (1973) Dilemmas in a general theory of planning. Policy Sci 4(2):155–169
53. Rockström J, Steffen W, Noone K, Persson Å, Chapin FS, III, Lambin E, Lenton TM, Scheffer M, Folke C, Schellnhuber HJ, et al (2009) Planetary boundaries: exploring the safe operating space for humanity. Ecol Soc 14(2):32
54. Semantix (2021) Data platform. Retrieved from Semantix.com.br. https://semantix.com.br/en/data-platform/
55. Shepherd DA, Sutcliffe KM (2011) Inductive top-down theorizing: a source of new theories of organization. Acad Manage Rev 36(2):361–380
56. Sprott JC (2003) Chaos and time-series analysis. Oxford University Press, New York
57. Stagich TP (2001) Collaborative leadership and global transformation. Global Leadership Resources
58. Sterman J (2002) System dynamics: systems thinking and modeling for a complex world. MIT Sloan School of Management, Cambridge. Retrieved from https://dspace.mit.edu/bitstream/handle/1721.1/102741/esd-wp-2003-01.13.pdf?sequence$=$1&isAllowed$=$y
59. Strobel G (2020) ESG scoring is failing: time for improvement. Retrieved from Forbes https://www.forbes.com/sites/forbesfinancecouncil/2020/07/02/esg-scoring-is-failing-time-for-improvement/?sh$=$51ffa68733da
60. SYSTEMA Erasmus (2022) The importance and potential of systems thinking and system dynamics in planning for sustainability. Retrieved from www.SystemaErasmus.eu. https://www.systemaerasmus.eu/wp/the-importance-and-potential-of-systems-thinking-and-system-dynamics-in-planning-for-sustainability/
61. Tanaka H (2019) Project & program management (upfront phase) in the digital transformation (dx) and green growth days, east asian practitioner's perspective. Eden Doctoral Seminar: Digital Transformation in Project Management. SKEMA Business School, Lille
62. Tapscott D, Williams AD (2008) Wikinomics: how mass collaboration changes everything. Penguin, London
63. Theraulaz G, Bonabeau E (1999) A brief history of stigmergy. Artif Life 5(2):97–116
64. Watts DJ (2004) The "new" science of networks. Annu Rev Sociol 30(1):243–270
65. Watts DJ, Strogatz SH (1998) Collective dynamics of 'small-world' networks. Nature 393(6684):440–442
66. Weick KE (1995) What theory is not, theorizing is. Admin Sci Quart 40(3):385–390
67. Zhang Q, Prouty C, Zimmerman JB, Mihelcic JR (2016) More than target 6.3: a systems approach to rethinking sustainable development goals in a resource-scarce world. Engineering 2(4):481–489

Chapter 4
From Concepts to Coherence: Modeling Social Impact Data for Societal Systemic Alignment

4.1 From Concepts to Coherence

The idea of societal automation is gaining traction. Integrating sustainability data into scalable, automated frameworks will enhance decision-making, and support policy design. The alignment of business activities with social benefits, known as "shared value," has also become a key focus for some stakeholders, including sustainable enterprises, social organizations, governments, and investors. The need for standardized, comparable data to effectively measure social impact presents a challenge in the dynamic societal context. However, the advancement of IT and digitization, particularly the proposed unifying model, holds promise for addressing this challenge through social automation leveraging collective intelligence.

4.2 The Scientific Approach: Assembling the Building Blocks

The Inductive Top down Theory Building Approach (ITDTA) by Shepherd et al. [7] was developed from coherence theory and holds pragmatist philosophical traditions with a problem-centered solution focus. ITDTA is generally an inductive approach as it draws from a wide range of information, resulting in a generalized theory. The term top-down may look like an error at first glance, but it is not. The deductive element, or top-down element, references the fact that existing empirical literature is the data applied for theory building. ITDTA applies abduction to address the real-world problem that the existing literature does not answer well. Deduction and abduction together detect the paradox from within the existing knowledge. While simultaneously, inductive and abductive reasoning develop a new theoretical model. It is an elegant iterative process guided by the researcher's existing knowledge and constrained by scholarly context.

© The Author(s) 2025   75
C. Aebi, *Unifying Sustainability Information for Societal Automation*,
SpringerBriefs in Business, https://doi.org/10.1007/978-3-031-83120-1_4

ITDTA offers a robust and rigorous theory-building method appropriate for understanding and resolving paradoxes. This approach was selected because it is effective at enhancing discovery within or across paradigms, and suitable when the existing literature is vast, dynamic, complex, and from disparate sources, as is present across the social value and social impact theme [7].

This methodological approach bridges the (hardware—physical reality) rigid positivist philosophical design required by computer algorithms and the (social construct—reality) constructivist philosophical design demanded by social scientists for social sustainability. The researcher is challenged to dance between both positivist and interpretivist philosophical awareness. ITDTA methodology can handle these contradictions for theory building that broadly unifies across specialist domains with problem-centered, solution-focused pragmatism. This dance takes both an ontological and epistemological stance on the world, theory, knowledge, and the relationship between knowledge and action to overcome rigor and relevance challenges by embracing both [9]. In essence, ITDTA operationalizes coherence theory, grounded in pragmatism. An iterative synthesis process produces the results to the extent that the theorist believes that the sensory and conceptual representations cohere and the explanations are superior to alternative explanations.

There is a challenge for a researcher to document this iterative process in a palatable manner for the reader. This book outlines the process using building blocks. Building Block I begins with the literature and is detailed in Chap. 2, focusing on data extracted from social value and social impact assessment literature. Patterns and relationships start to take shape as a sensory representation, including mapping and requirements gathering. Building Block II involves contextualization with thought trials through metaphor, covered in Chap. 3. Building Block II further evaluated the potential contribution and identified practical components to be assembled as conceptual representations. However, it should be noted that these building blocks were not independently formed in one step by step process. The cyclical repetition and reviewing of each building block was repeated through several iterations until saturation of ideas and refinement of a new theory were achieved. The new concept was tested via scientific experimentation and is claimed to be valid only if useful [5].

4.3 The Model

The Stigmergic Information System Network for Social Sustainability conceptual model is a guide to support the development of technologies that will interact, operate, and think in line with society's expectations in terms of social value. This model leverages the new phenomenon of scale-free distributed technology for social sustainability. It is structured to offer predictive and suggestive analysis to minimize negative social impact. Stigmergic self-organization processes for data capture and information aggregation leverage collective intelligence to guide decision-makers toward activities that solve social problems.

The Stigmergic Information System Network model conceptualizes the patterns that connect different social sustainability indicators for systemic alignment with social value. It offers a structure to address the unresolved technical big-picture issues by providing a stable minimal foundation structure that considers the dilemma of dynamic societal complexity within the paradox of eternal change.

The model is presented with descriptions of the following:

- System scales of participation
- Overview of the Stigmergic Information System Network for Social Sustainability
- How it works: systemic alignment with social value for social sustainability

The conceptual model offers a foundational structure for technology to support society and evolve with change. This conceptual model is a skeleton framework, which should, theoretically, function under the rules and order with which a society agrees. It forms a functional system based on the participant's inputs with fundamental algorithmic hierarchies to support decision-making within boundary conditions as a social sustainability thermostat. The pragmatic solution works within the systems we have today with the flexibility to adapt to new structures in the future.

The Stigmergic Information System Network for Social Sustainability is designed to capture, measure, and connect the information to support decision-making and cross-disciplinary data integration for socially relevant data. Computational modeling uses computers to study complex systems. It allows for adjusting variables within the system for simulation and observable outcomes. The ecosystem consists of semantic platforms with data loaders, sorting, labeling, linking, machine learning, dashboards for visualizations, and broader social information network sharing across hubs for information aggregation.

4.3.1 System Scales of Participation

A systems view provides a perspective for understanding society's existing relationships and processes, and its participants activities and will to endure. *Society* is a system constantly renewing and redefining itself with no end. Across society, autonomous systems operate/live within other autonomous systems, operating and living within even larger autonomous systems.

The model acknowledges the following factors regarding all living systems and their relationship to other living systems in their environment [6]. Each living system has the characteristics of autonomy, circularity, and self-reference. These characteristics afford the ability to self-create and self-renew. Living systems aim to (re-) produce themselves. The identity of the living system is its essential product. Living systems have "environments," but the relationship with their environment is internally determined. A system's interaction with its environment is a circular interaction process exclusively to facilitate self-production.

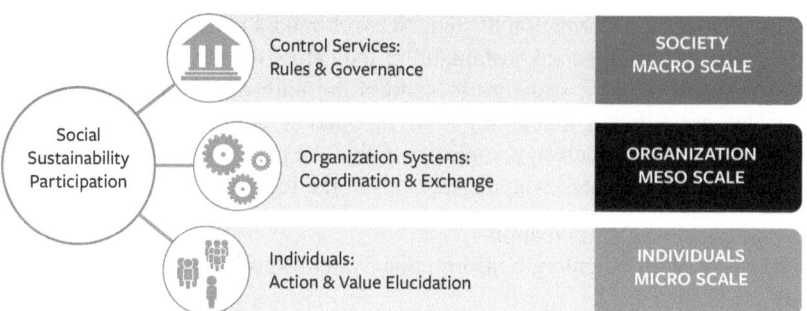

Fig. 4.1 Scales of participation for stigmergic information system network for social sustainability

Defining scales allows for classification that supports structuring architectures, communication, algorithms, and governance while retaining the flexibility needed for an ever-changing society. Figure 4.1 presents the scales of participation for this conceptual model.

Macro-scale is society participating with rules, control services, governance, and risk monitoring. It is the most extensive autonomous system with entrenched democracy, capitalism, religion, and human rights systems at a geographic, political, or ideological level.

Meso-scale is the level of an organization participating in coordination and resource exchange. Autonomous systems operate/live within other autonomous systems, operating and residing within even larger autonomous systems such as groups and groups of groups in the form of for-profit and not-for-profit organizations, associations, communities, and sports clubs. Business is a form of organization that is also a social system.

Micro-scale is the individual's participation and actions, e.g., production and consumption, and value elucidation, e.g., goal values and social priorities. Each individual is a living system characterized by autonomy, circularity, and self-reference with the ability to self-create and self-renew.

All participants can use anonymized data, potentially in real-time, for decision-making. The data is first cleaned to remove confidential information. Metadata is open and shared with relevant parties for the social sustainability ecosystem, including the comparability of data against business and market competitors.

Individuals and groups hold unique perspectives within the system. Role labeling follows the definitions applied in the SSM methodology CATWOE. Role distinctions apply to all micro, meso, or macro scales. A single individual may quantify more than one role perspective for an activity. Quantifying social value/social impact from various individual perspectives at the micro-scale is a new opportunity.

4.3.2 Overview of the Stigmergic Information System Network for Social Sustainability

The foundational structure is a minimum set of criteria that facilitates societal automation to be supported with data that supports social stability within dynamic processes that leverage collective intelligence. A minimal set of adaptive and flexible structures, hierarchies, and controls provides room for the systemic organization to manage through a continually changing context.

The model overview provides perceptual wholes that distinguish scales and relationships of information flows between these scales within the overall system. It diagrams how the data structures influence systematic behaviors. Quantification of social value is captured and shared between individuals and groups through a stigmergic information system (SIS) ecosystem.

Human short-term, long-term, and existential goal value models for artificial agents and algorithms are out of the scope of this research. However, the Complex Controller Structure described in Chap. 3 is a structural element for quantifying micro-scale goal values at the individual micro scale. As described in Chap. 3, existential goal values of Formal Axiology are an introductory example that passed model review. Theories for human motivation, and ethical decision models can be applied to short-term and long-term goal value structures as a drop-down selection based on user preference.

Figure 4.2 provides an overview of the Stigmergic Information System Network for Social Sustainability. It indicates the fundamental interactions and connections between the scales within the system. It diagrams how the structures influence systematic behaviors.

The conceptual model includes a prototyping and testing arena separate from active algorithms but integrates all scales. Policymakers can receive guidance with analytical, diagnostic, and prescriptive insights from the network data, consultation with expert groups and academia for prototyping, and public engagement for input and testing models. Due to scale, the processes of operational engagement with individuals are supported by automation. The arrow from prototyping to the SIS ecosystem is indicative that operational tasks, including public engagement, are executed from within the SIS ecosystem.

The Stigmergic Information System Network for Social Sustainability Conceptual Model has a scale-free base operational architecture. Today's digital ecosystems consist of multi-agent distributed information processing systems across a distributed data mesh logical architecture connected to semantic platforms for sorting and sharing social value data across decentralized networks. Architectures are rapidly evolving with artificial intelligence and quantum computing. Technology mediates the measurement and provides analysis of social aspects of activities. Technology initiates micro-tasks of self-organization processes programmed with swarm intelligence methods. As described in Chap. 3, cyber-physical coordination through stigmergic meta-processes triggers actor actions in the environment.

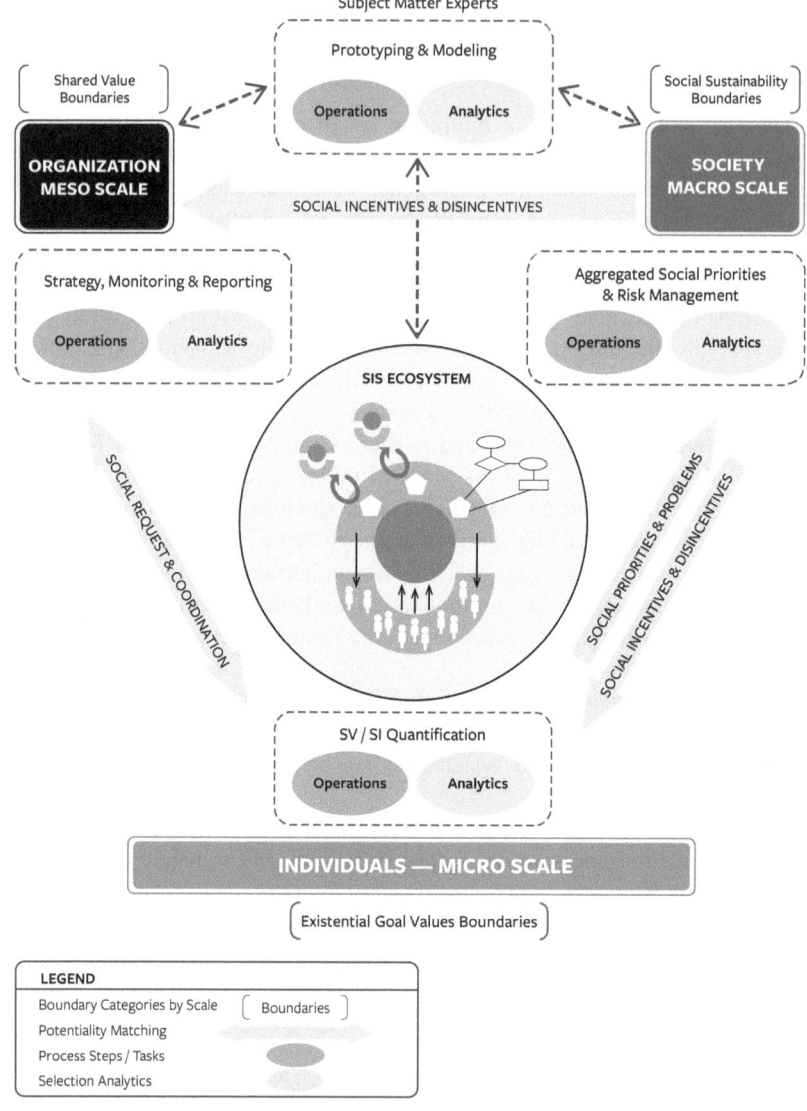

Fig. 4.2 The stigmergic information system network for social sustainability conceptual model

4.3.3   How it Works: Systemic Alignment with Social Value for Social Sustainability

Continuous transformation and emergent order are natural. Learning loops allow for ongoing refinement and adaptation of modeling. Automation of operational tasks for data gathering leads to efficient scalability.

Aristotle pointed out that humans interact within society, and different self-interests are always at play. Accommodating different interests is a concern of a coherent society and traditionally a responsibility within the governance role. An information system network for social sustainability includes a base structure that is open to the interests of all and capable of containing destructive power plays.

Changes within social systems are complex and nonlinear. Each change creates a repercussion or unique effect on other problems. Actions taken to address one problem may lead to unintended consequences for another problem.

System dynamic modeling provides the tools for simulation and predictions to support decision-making. Modeling with positive and negative feedback loops programmed into simulations can uncover unintended consequences, predict system reactions, and foresee delayed benefits for better planning.

Applying dynamic system modeling offers a structure for incorporating feedback loops of interaction into analytics to provide insights about the dynamics of the social systems. Dynamic feedback loops offer predictive, corrective, and retrospective analytics with learning value control loops applied against control module parameters to act as systemic thermostats for stability, reduce perception bias, and monitor exploitative effects.

Figure 4.3 shows the learning feedback loops for dialectical operational process automation and analytics within the Stigmergic Information System Network for Social Sustainability Conceptual Model.

Screening establishes a baseline and develops models to estimate future impact. Dialectic analysis provides a lens to visualize and understand the effects of change in our past, present, and future. Macro-scale models are built using the soft systems

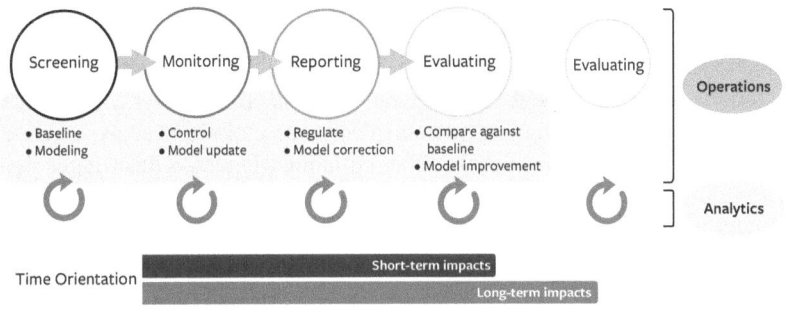

Fig. 4.3 Learning feedback loops for the dialectical operational processes and analytics within the stigmergic information system network for social sustainability

methodology *Macro Scale Modelling Process for Societal Value in Change Activity* [3]. This process engages and synergizes participant expectations and actions. System archetypes are foundational patterns for model building, as described in system archetypes for system dynamic theory [2].

Feedback loop controls monitor for disturbances within a system. The automation of minimal rules and control services in the form of boundary conditions can offer decision-making options to pull a system into states of equilibrium (or near equilibrium). Monitoring provides the opportunity to mitigate unpredictable effects and impacts. Reporting compares results against model expectations, and these results are applied to model corrections. Evaluation over time considers short-term and long-term results that further feed into model improvements.

Nonlinear complex systems are difficult to analyze. The evaluation process also provides a corrective system design. Understanding and defining the existence of one state requires defining what is not within the scope of that state. Experimentation with a model allows for adjusting parameters to study the differences in outcomes through computational experiments. Such simulations can be valuable predictive agents in organizations from a societal or individual scale.

Although governments and institutions seek transparency in business activities, there remain open questions and exploitation risks. When algorithms are open and transparent for automated decision-making, those who wish to exploit these algorithms gain a deeper understanding of how to manipulate them. Exploitation falls into the category of social value destruction for the individual and risk management for society. Learning feedback loops provide corrective suggestions after the fact. Rules or control services that do not serve a stabilizing equilibrium or result in destabilization are monitored and highlighted for review, reassessment, redesign, or decommissioning.

Technology-mediated social system projects operate under different influences from the current diplomatic and competitive systems. There lies a logical relation, communication, and interaction; a ubiquity of emergence ensures autonomy yet synergizes harmony and equilibrium instead of brute force interactions.

Structuring social participation for a minimum set of criteria that facilitate the flowing process of change allows for self-organization rather than predesigning and expecting to control the outcomes. Micro, Meso, and Macro scales cooperate for dynamic signaling action, modeled as semiotic chains for communication automation and computational support in analysis and decision-making suggestions. Structurally, this involves separating operational tasks from analytical information.

Semiosis is part of an ongoing process of communication and analysis tasks. Any change registered in the system sets off an initiating trigger. As this trigger signal is given, it initiates an interpretation or condition at a focal level. This interpretation or condition is processed within limitations or boundary conditions. Existential goal values, shared value, social value, social impact, and social sustainability are quantified, computed, and presented in perspective relevant to the viewer at various scales.

Machine learning methods programmed with decision trees are applied as selection criteria for analysis to ensure social sustainability by promoting equilibrium and stable states. Cybernetics influence the practices that create problems to motivate change within sustainable boundaries. Note that random disturbances can create unpredictable impacts throughout the system. Control services guide the system into equilibrium states (or near equilibrium). A systemic gravity pushes or pulls spontaneous self-organization into form despite the unpredictability.

Citizen and customer engagement fixed into information system design for machine learning reduces perceptual distortions and synergizes solutions for an enduring society.

4.4 Conclusion

The proposed conceptual model for the Stigmergic Information System Network for Social Sustainability aligns with the broader context of unifying sustainability information for societal automation and leveraging collective intelligence for a sustainable future. By providing a stable and minimal foundation structure, this model addresses the complex and dynamic nature of societal systems. It offers a framework for connecting information to support decision-making and cross-disciplinary data integration, aiming to maximize social benefits and minimize negative impacts.

The conceptual model:

- Defines scales and participation levels
- Diagrams the structures and information flows that influence systematic behaviors
- Sets boundary conditions at each participation scale for programable parameters technology can apply as selection criteria and analysis for social sustainability
- Expresses the fundamental process interactions and connections for stigmergic scale-free self-organization
- Diagrams the learning feedback loops for dialectical operational process automation and analytics

The model promotes collective intelligence and facilitates a systemic approach to addressing social demands and problems by emphasizing transparency, comparability, and sharing of social sustainability data. Furthermore, its predictive, corrective, and retrospective analytics, along with feedback loops for social stability, align to leverage technology to drive sustainable practices and societal automation. The model's focus on self-organizing communication micro-processes and increased transparency also contributes to the overarching aim of leveraging collective intelligence for a sustainable future by fostering collaboration and knowledge sharing across diverse sectors and disciplines.

4.4.1 Research Implications

Such topics require the researcher to balance a spectrum between academic rigor and managerial relevance. This conceptual modeling advances the practical discussions for technology development to facilitate data capture, sharing, coordination, and automated communication for social responsibility for developers, computer scientists, social sustainability practitioners, business managers, and policymakers.

The conceptual model incorporates the new phenomenon of scale-free distributed technologies for a technically feasible elucidation to bring together systemic synergies and reduce the complexity and costs associated with assessing social value.

Understanding the direct and indirect social impacts across roles over time is a strategic opportunity for product and service development, risk management, and financial management.

A future implemented stigmergic information system network for social sustainability would offer businesses an increased opportunity to align their activities with social value.

Managerial recommendations offer practical actions businesses can take to prepare and align to capture these advantages.

This conceptual model offers a skeleton concept. This base concept is both complete and incomplete. It offers a unifying minimum set of stable mechanisms for model development and platform growth across information structures. Continual evolutions of rules, organizations, groupings, and boundary conditions can be implemented without fundamentally disrupting the minimal structure.

Boundary limitations are drawn into the model to describe the computational structure supporting decision-making. Although Social Sustainability, Shared Value, and Existential goal values are set as boundary limitations at societal scales, the model provides only examples from existing knowledge of these criteria. Defining the exact minimum set criteria of boundary conditions of social sustainability, shared value, and individual existential goal values is outside the scope of this research.

Additional factors that are out of the scope of this research:

- Definitive software engineering and architecture
- Details of human goal-value models for artificial agents
- Decision algorithms and predictive modeling
- Process diagramming for workflow automation

Any single theoretical paper on a new phenomenon is not an end. It is merely a tiny step in an ongoing, dynamic process along a continuum to represent "interim struggles" that may help theorists inch toward increasingly coherent and refined theories [10]. It is recommended but not clear that academia would be the party to advance and refine this model before an ecosystem is built. However, businesses could develop the information system network for social sustainability, and academia will refine models while policymakers apply rules and regulations.

4.4.2 Further Studies

A third-party review, improvement, and validation of the generalizability of this model are recommended. Developing information models using machine learning and algorithm models would further this concept. Software development and architecture engineering that constitute the conceptual model network and ecosystem further advance the practical refinement of this model. The elaboration and harmonization of Tier 2 classifications, categories, labeling, characteristics, and types across assessment methods and roles will be necessary if this model is further advanced. Additional research targeting the automation of data gathering of social sustainability, social responsibility assessment methods for cost efficiency, and improved data consistency is also recommended. Finally, Axiology-based human goal-value models for artificial agents and algorithms for quantifying social value options are valuable research that can advance the testing and application of this model.

4.4.3 Discussion

During this research, a question was asked: Is individual input necessary, relevant, or accurate? After all, human values are contradictory, changeable, manipulatable, implicit, and under-defined [1]. Each individual is a living system characterized by autonomy, circularity, and self-reference with the ability to self-create and self-renew. Social sustainability identifies and manages positive and negative impacts on people [8]. Further, all systems operate with a unique perception and perception bias. Organizations demonstrate perception bias when measuring social sustainability [4].

Furthermore, governments have a different perspective, focusing on decision-making to reduce risks. The individual brings the perspective of environmental pressures and the motivation to thrive for themselves and those who are important to them. The aggregated information drawn from the pressures and motivations offers insights toward policy with new visibility on social priorities to align and guide activities and social risk mitigation. The feedback loop to and from the individual indicates if predictions were achieved or not achieved and aids in learning for all parties at all scales.

A recurring discussion point was the conceptual model's overarching purpose. There were discussions about whether the model should have the purpose of systemic alignment with social value, social responsibility, or social sustainability. Although not the overarching purpose and global boundary condition, all three dimensions are included in the model. Social responsibility heavily impacts the business perspective and does not represent society as a whole. However, the boundary condition of Shared Value includes social responsibility.

Systemic alignment with social value makes a clear positive drive for society to advance and thrive. Algorithms with the model's boundary conditions should theoretically offer suggestions harmoniously with social value. However, the ability to advance all individuals and groups across all sectors to social value seemed too big a first step. It also risked the dynamic flexibility society appreciates as autonomy without constraints against activities that do not deliver the highest social value.

Setting the purpose of social sustainability allows room for the self-determination of action and choice within the boundaries that avoid harm to others. It leaves room for choice, creativity, and trial and error. It is less restrictive and more tangible for programming and rewards motivation for groups and individuals. This is not to say that group, community, or national goals cannot be defined and worked towards, bringing society to a state of thriving or the most significant potential for the broadest social value within this model. On the contrary, one new solid contribution from this conceptual model is that it includes the goal values of individuals or groups, including community aggregation. Further, organizations and policymakers would gain a new data set for alignment with social priorities and the communication channel to build society-level aspirational goals.

The question of net shared value as a Meso boundary condition also arose. Can we be sure that it is possible to result in net shared value for an organization consistently? For one group to get ahead, is not another left behind? Shared value proposes that business has a high-synergy role for collaboration. Gathering data sets of social aspects for computer-aided suggestions for decision-making and reporting will help us answer the question. Learning loops and improved models for calculating the social benefits and harm of activities across society should provide visibility into the imbalances and destabilizing activities.

Another discussion point that came up frequently was whether it is possible to define an accurate baseline in an ever-changing, highly complex world. A fixed "start" may not always be possible; however, snapshots and trends as points for algorithms and models should be possible.

A criticism of the soft system methodology (SSM) applied to decision-making for social sustainability is that it assumes the status quo and seeks consensus. Political issues of unequal power distribution, resource scarcity, and media distortions could lead to technology locking in a system that society or communities may one day wish to transition. This method considers consensus in model building, which could serve the interests of currently dominant parties. Consensus would serve the interests of dominant parties of ownership and existing rules and control systems.

This research acknowledges this criticism of SSM. Real-time self-organization is a tricky balance and an entirely new means of social order, with implications for democracy and market dynamics. The development of Macro-scale models is designed to incorporate the feedback of all parties involved. Models are designed to highlight where existing rules and models are destabilizing, bringing data-driven insights into power structures' imbalances. It is a new opportunity for those parties who may be negatively impacted to become involved in the early discussions and to have a voice in modeling and testing.

The model is not designed to make decisions like an AI for an imaginary society. Any idealistic notions within the research represent ideals and societal goals presented within the existing literature. There are no political or political system restrictions within the modeling. Further, there is no underlying proposition of Authoritarianism, Fascism, Technocracy, Democracy, Capitalism, Communism, Marxism, or any "ism." The model is fundamentally designed as a foundational logical structure that can be applied to stigmergic information systems for social sustainability regardless of and within any legal order as programmed within the control structure modules.

It offers an initial step. It considers a potential future in which human and cyber interactions introduce a mechanism of algorithms that seek systemic equilibrium and alignment with social value for social sustainability.

According to the pragmatic philosophy, this research applied a problem-centered solution focus. This conceptual model is tested via scientific experimentation and claims to be valid *only* if useful [5].

References

1. Armstrong S (2015) Motivated value selection for artificial agents. In: Workshops at the twenty-ninth AAAI conference on artificial intelligence
2. Braun W (2002) The system archetypes: the systems modelling workbook. https://www.albany.edu/faculty/gpr/PAD724/724WebArticles/sys_archetypes.pdf
3. Checkland P, Poulter J (2020) Soft systems methodology. In: Systems approaches to making change: a practical guide. Springer, Berlin, pp 201–253
4. Doyle TM (2018) Ratings that don't rate: the subjective world of ESG ratings agencies. American Council for Capital Formation, pp 65–71
5. Legg C, Hookway C (2019). Pragmatism (E. N. Zalta, Ed.) Metaphysics Research Lab, Stanford University. Retrieved from Stanford Encyclopedia of Philosophy. https://plato.stanford.edu/entries/pragmatism/
6. Maturana HR, Varela FJ (1991) Autopoiesis and cognition: the realization of the living. Springer, Berlin
7. Shepherd DA, Sutcliffe KM (2011) Inductive top-down theorizing: a source of new theories of organization. Acad Manag Rev 36(2):361–380
8. United Nations (2022) Social sustainability. Retrieved from United Nations Global Impact. https://www.unglobalcompact.org/what-is-gc/our-work/social
9. Vo LC, Mounoud E, Rose J (2012) Dealing with the opposition of rigor and relevance from dewey's pragmatist perspective. Management 15(4):368–390
10. Weick KE (1995) What theory is not, theorizing is. Admin Sci Quart 40(3):385–390

Glossary

Actual social impact The resulting social impacts experienced by communities due to an event, project, or change, not predictions or forecasts.

Artificial Intelligence Intelligence and decision-making capability of machines that is similar to the natural intelligence displayed by humans.

Benchmark A comparative point of reference point or range. The social indicator is set as a standard or norm for monitoring value.

Civil society The network of individuals and groups (both formal and informal)—and their connections, social norms, and practices—comprises a society's activities that are separate from its state and market institutions. It includes religious organizations, community groups, foundations, guilds, professional associations, labor unions, academic institutions, media, advocacy or pressure groups, political parties, etc.

Collective Intelligence Collective intelligence is the enhanced problem-solving and decision-making ability that emerges from collaboration and data sharing among a group of people. Collective intelligence leverages the diverse perspectives, skills, and information of a group to achieve outcomes that would be difficult or impossible to achieve individually. The integration of human and machine contributions often enhances this process, leading to more robust and effective solutions.

Community A concept generally refers to a place-based grouping of people with some sense of shared identity, interactions within everyday life, and some joint social and political institutions.

Complex Controller Structure A conceptual structure for human computation illustrates aspects of human reasoning with external looping for modeling decisions and internal looping for goal-value decisions.

Computation Information processing.

© The Author(s) 2025
C. Aebi, *Unifying Sustainability Information for Societal Automation*,
SpringerBriefs in Business, https://doi.org/10.1007/978-3-031-83120-1

Computation model A conceptual model for computation to interpret the behavior of a complex system.

Conceptual model A diagram representation of an abstract idea or system, made of the composition of concepts that are used to aid in understanding Corporate Social Responsibility. The responsibility of enterprises for their impact on society.

Crowdsourcing An aggregation of the intellectual skills of a large number of people for a web/based project.

Cybernetics Theory of control in technical, biological, and social systems. Management pursues and sustains a defined goal, e.g., a physical state against a changing environment.

Data mesh An architectural concept that supports analytical data at scale, offering access to a constantly changing number of distributed domain data sets for a proliferation of use scenarios, including machine learning, analytics, or data-intensive applications required for future AI.

Dilemma a situation in which a difficult choice has to be made between two or more alternatives, especially equally undesirable ones.

Distributed Cognition The use of information technologies to make distributed information processing by humans much more powerful, focused and efficient Distributed thinking systems Systems created of many thinking technologies capable of complex and sophisticated problem-solving capabilities.

Emergent human computation A system where the natural consequences of individual and collective behaviors are observed to inform a more profound understanding in the context of system dynamics.

Engineered human computation A system explicitly defines an interaction context or predictive decision action to produce a desired systemic result.

Environmental Impact Assessment (EIA) A formal assessment process for the prediction of likely environmental consequences (positive and negative) of a plan, policy, program, or project before implementation, usually as part of the regulatory (environmental licensing) requirement.

Framework A supporting structure around which something can be built Global reporting initiative (GRI) An international, independent non-profit organization established in 1997 in partnership with the United Nations Environment Program (UNEP), which promotes economic, environmental, and social sustainability.

Good governance A normative understanding of how governance (of any organization) should be performed. For example, accountability, transparency, the rule of law, inclusion, and a participatory process.

Human Computation Modeling Techniques for modeling human participation with large-scale thinking systems for purposeful social decision-making.

International Finance Corporation (IFC) The International Finance Corporation is the World Bank Group's private sector lending arm. Its performance standards are the basis of the Equator Principles, which are increasingly an international benchmark.

Impact An economic, social, environmental, and other consequence that can be reasonably foreseen and measured in advance if a proposed action is implemented.

Impact assessment The process of identifying the future consequences of the current or proposed action.

Impact History This refers to the experience a community has had with other projects. This affects how they relate to new projects and how much trust they might have. It also means that there may be legacy issues that an operator has to deal with.

Indirect impact The impact that occurs due to an indirect change caused by a planned intervention. For example, a physical change to the environment: a mine may increase river turbulence resulting in a reduction in the supply of fish which may negatively impact the economic livelihoods.

Machine learning The use and development of computer systems capable of learning and adapting through algorithms and statistical models to draw inferences and analyze patterns in data.

Needs assessment A systematic procedure for determining individual or community issues, including ranking importance as a component of public development programs.

Nested systems Systems that encompass other systems and are also enclosed by even other systems.

Node A device or data point in a more extensive network.

Paradox a statement that is seemingly contradictory or opposed to common sense and yet is perhaps true.

Shared value A recognition that an organization must create both social benefits and benefits for the organization. This view also acknowledges that social harms frequently create costs in the form of social risks and therefore need to be carefully managed.

Social impact The portion of the total outcome that is experienced or felt, in a perceptual or corporeal sense at the level of an individual, social unit (family/household/collectivity), or community/society due to an event, project, or change above and beyond what would have happened anyway.

Social responsibility In addition to maximizing shareholder value, businesses must act to the benefit of society.

Social return on investment (SROI) A method for measuring the social benefit beyond immediate financial return investment for projects or activities. They are typically represented as a ratio relative to resources invested.

Social Sustainability This research defines sustainability as the capacity to endure. Moreover, the term social sustainability applies this definition to the capacity of human society to endure.

Social value The quantification of the relative importance experienced or felt, in a perceptual or corporeal sense at the level of an individual, social unit (family/household/collectivity), or community/society.

Societal automation The integration of automated systems and technologies into various societal functions to enhance efficiency, improve outcomes, and create synergistic benefits across daily life and industries.

Stakeholders Individuals and groups affected by or that can affect a project or operation. Stakeholders may consist of individuals, interest groups, organizations, and institutions.

Stigmergy Stigmergy is a mechanism of collaborative indirect coordination or self-organization between agents or actions within an environment.

Sustainable development Development that meets the needs of the present without compromising the ability of future generations to meet their own needs.

System dynamics An approach to understanding the non-linear behavior of complex systems over time using stocks, flows, internal feedback loops, table functions, and time delays.

Systems thinking A set of synergistic analytic skills is used to improve the capability of identifying and understanding systems, predicting their behaviors, and devising modifications to them to produce desired effects. These skills work together as a system.

UN Global Platform A cloud-service ecosystem to support international collaboration to develop Official Statistics using new data sources and innovative methods and to help countries measure the Sustainable Development Goals (SDGs) to deliver the 2030 Sustainable Development Agenda (UN Committee of Experts on Big Data and Data Science for Official Statistics).

Unpaid work Labor that does not receive a direct wage. This production of goods or services may be consumed inside or outside a household. The work is distinguished from leisure if a third person could be paid to perform the activity. This work is not included in the calculation of the System of National Accounts and Gross Domestic Product economic figures. Examples of unpaid work are domestic labor, charity work, and unpaid employment.

Values The individuals' abstract and often subconscious assumptions about what is proper and important organized into a value system that can vary substantially between cultural groups.